THE
ORBITAL PERSPECTIVE

On the cover is a photo that my crewmate and space walking partner Mike Fossum took of me while I was in the cupola of the International Space Station. The photo was taken from the Russian docking compartment Pirs as we passed over coastal Australia.

I have always called this picture *Downside Up Down Under*. Peter Gabriel inspired the name of the photo through the many discussions he and I have had, both on Earth and while I was in space, about the nexus between space, music, culture, and the orbital perspective. Peter's song "Downside Up" and his explanation of the song perfectly capture the sentiment of the photo and the essence of the orbital perspective: "At one point the world turns upside down, and you have to imagine lying on a field of grass and looking up at the sky long enough that you start to see the sky as down, as an ocean below you, and that is the state 'downside up.'"

> *Downside up, upside down*
> *Take my weight from the ground*
> *Falling deep in the sky*
> *Slipping in the unknown*
> *All the strangers look like family*
> *All the family looks so strange*
> *The only constant I am sure of*
> *Is this accelerating rate of change*
>
> . . .
>
> *I stand here*
> *Watch you spinning*
> *Until I am drawn in*
> *A centripetal force*
> *You pull me in*
> —PETER GABRIEL

THE
ORBITAL PERSPECTIVE

Lessons in Seeing the Big Picture from a Journey of Seventy-One Million Miles

ASTRONAUT
RON GARAN

BK

Berrett–Koehler Publishers, Inc.
a BK Currents book

Berrett-Koehler Publishers, Inc.
1333 Broadway, Suite 1000
Oakland, CA 94612-1921
Tel: (510) 817-2277   Fax: (510) 817-2278   www.bkconnection.com

ORDERING INFORMATION
Quantity sales. Special discounts are available on quantity purchases by corporations, associations, and others. For details, contact the "Special Sales Department" at the Berrett-Koehler address above.
Individual sales. Berrett-Koehler publications are available through most bookstores. They can also be ordered directly from Berrett-Koehler: Tel: (800) 929-2929; Fax: (802) 864-7626; www.bkconnection.com.
Orders for college textbook/course adoption use. Please contact Berrett-Koehler: Tel: (800) 929-2929; Fax: (802) 864-7626.
Orders by U.S. trade bookstores and wholesalers. Please contact Ingram Publisher Services: Tel: (800) 509-4887; Fax: (800) 838-1149; E-mail: customer.service@ingram publisherservices.com; or visit www.ingrampublisherservices.com/Ordering for details about electronic ordering.

Berrett-Koehler and the BK logo are registered trademarks of Berrett-Koehler Publishers, Inc.

Printed in the United States of America

Berrett-Koehler books are printed on long-lasting acid-free paper. When it is available, we choose paper that has been manufactured by environmentally responsible processes. These may include using trees grown in sustainable forests, incorporating recycled paper, minimizing chlorine in bleaching, or recycling the energy produced at the paper mill.

Library of Congress Cataloging-in-Publication Data
Garan, Ron (Ronald J.), Jr., 1961–
    The orbital perspective : lessons in seeing the big picture from a journey of seventy-one million miles / astronaut Ron Garan.
      pages cm. — (A BK currents book)
    Includes bibliographical references and index.
    ISBN 978-1-62656-246-2 (hardcover)
    1. Garan, Ron (Ronald J.), Jr., 1961– 2. Astronauts—United States—Biography.
3. International Space Station. 4. Astronautics—International cooperation. 5. Peaceful change (International relations) I. Title.
    TL789.85.G37G37 2015
    629.450092—dc23   2014038722
    [B]

First Edition
20 19 18 17 16 15 | 10 9 8 7 6 5 4 3 2 1

Produced and designed by BookMatters; copyedited by Todd Manza; indexed by Leonard Rosenbaum; proofed by Janet Blake. Cover design by Steve Pisano. Front cover photo by Mike Fossum and reproduced courtesy of NASA.

To my family: my wife, Carmel, and our sons, Ronnie, Joseph, and Jake. Thank you for being my inspiration and for enabling me and accompanying me on this incredible journey. And to all those laboring to make life on Earth as beautiful as our planet looks from space.

CONTENTS

FOREWORD

As you read of Ron's discovery of his life's goal and purpose during his time in space, I hope you will also reflect on your own life's goal and purpose here on Earth. What are the problems you consider impossible to solve—hunger, poverty, war, environmental issues—and how can you gather together with others to solve these problems? Use Ron's idea of the orbital perspective as a way to erase obstacles, boundaries, and resistance to any problem.

If you had asked bankers thirty years ago whether a bank could lend billions of dollars to millions of extremely poor women in a country like Bangladesh without any collateral and with nearly a one hundred percent repayment rate, they would have laughed at you! Today, this is done globally—even in New York, San Francisco, and Los Angeles—and is a tested tool to counter poverty and give self-reliance. Another kind of business, a nondividend business, is becoming popular globally to address problems that were never considered amenable to sustainable solutions by business methods. What other tools can you discover if you put your mind and heart into the question?

Ron has shared with us his unlimited view of Earth's possibilities from the vantage point of the International Space Station. May we use this beautiful perspective for the good of all with whom we share our beautiful home, this fragile oasis known as Earth.

Muhammad Yunus
Nobel Peace Laureate

PREFACE

I have wanted to write this book since returning to Earth from my first space mission in 2008. Launching on Space Shuttle *Discovery* fulfilled of a lifelong dream of flying in space. It also marked the start of another quest—a quest that would lead me to reject the status quo on our planet, a quest to help reduce the sobering contradiction we see when we look at our planet from space, and a quest to help make life on Earth as beautiful as the visible beauty of this planet when seen from space.

I returned from that mission with a compelling need to share a profound feeling of hope that was singed into my awareness from the experience. I call this awareness the orbital perspective, and this awareness came with a responsibility to highlight the need for effective global collaboration. My goal in writing this book is to spark change—very simple, but exponentially powerful change. I was compelled by a simple idea: our world still faces numerous problems because we have not, to date, learned how to work together. But I believe that the entire landscape of our society is changing, making available powerful collaborative tools that can engage the collective genius of our global society like never before. We are living in times of unprecedented interconnectedness—an interconnectedness that empowered a fruit vendor in Tunisia to spark a collective movement that has changed the world.

We live in a world where exponentially increasing technological advancements and interconnectivity are making the "impossible"

possible on a daily basis. Commonly held beliefs about what's possible and what's not are being overcome by events. We do not have to accept that the suffering and conflict on our planet are inescapable. I believe we live in a world where the possibilities are endless and limited only by our imagination and our will to act. The secret to achieving these critical things is to pull back to the point where we can see all the pieces of the puzzle and how they fit together, and where it also becomes apparent that our own backyard is bigger than we think. Our sphere of influence is in fact global. This is the orbital perspective.

I wrote this book to help create a global movement—a movement of inhabitants of this planet, known as humans, who believe that nothing is impossible, who are willing to set aside their differences and work together toward our common goals. I wrote this book to start a global conversation and to spread the awareness that it is no longer acceptable to have a go-it-alone attitude. The problems facing our world are too big, the challenges too critical, and the potential catastrophes too imminent to adhere to the old way of doing things.

We are past the point where any one organization or nation can go it alone and still effect any kind of real course correction to the trajectory of our shared future. The missing link that would bridge the gap between our problems and their solutions is simple to state yet challenging to implement. The answer is collaboration—true collaboration, which I will define and describe throughout this book, not the false collaboration that has become the latest buzzword.

There may be a tendency to discredit my words as overly optimistic or unrealistic. But I understand that, in the end, all governments will make decisions that are in accordance with what they believe to be in their national interest, and that many perverse incentives are stifling progress toward achievement of our goals. My view comes from seeing the worst and best we as a human species have to offer, from visiting the heights of space to working with those in destitute poverty. During my time in combat, I was routinely shot at and experienced the death of squadron mates and friends. I have experienced firsthand and have participated in the horrors that partly define life on this planet. Nonetheless, I remain optimistic that we can rise above the challenges we face.

FRAGILE OASIS

I hope that you find this book entertaining, engaging, and thought-provoking, but if after reading it all you have received is entertainment and a slight shift in perspective, I will have failed. My intention is to communicate a call to action. I am asking everyone to look for ways to create exponential, disruptive, positive action—action that leads to exponential advancement toward solving the challenges facing our world. I do not claim to have all the answers, and I know that there is great complexity in the topics I discuss, but I hope that after reading this book you will agree that we're on to something big that can potentially change the present trajectory of our global society and put it on a profoundly more positive path.

During and after my first mission, I was somewhat frustrated that I couldn't share my amazing experience with people as it was happening. So when I next was assigned to a six-month long-duration mission, I looked for a way to share this experience. I wanted to bring people along not just as spectators but as fellow crewmates.

What evolved from that desire is an effort called Fragile Oasis. Initiated with the help of the amazing Elyse David and then supported by NASA's Beth Beck and others, Fragile Oasis took shape and was launched prior to my mission in 2011. The goal of Fragile Oasis is to use the orbital perspective to inspire people here on Earth to make a difference. As a step in that direction, we have built collaborative tools into the site, which people and organizations can use to work together. We also wanted to identify fellow crewmates on "Spaceship Earth" who are making a difference.

Periodically, we award Fragile Oasis medals, which flew with me in space for more than 65 million miles, in the categories of Peace, Humanitarian Service, Education, and Environment Stewardship.

Fragile Oasis is envisioned as a vehicle to effect real change. We want to provide a means for people and organizations to collaborate and develop synergy toward overcoming our planet's challenges. We want to encourage people to make a difference and help organizations striving to make the world a better place to reach their goals. Fragile Oasis is still a work in progress, but I am really excited for its potential.

I hope you will tell us about what you're doing to make a difference. Register your projects on FragileOasis.org so we can help support your efforts. Visit OrbitalPerspective.com, so we can help get the word out. Share your ideas on how we can better work together on UnityNode.org, so that we can better coordinate the efforts of all.

Solving the challenges affecting our planet really depends on how each of us individually responds to them. I am asking you to join us in a global conversation centered on taking a big picture and long-term view. I am asking that you look for ways to combine what you and your organizations, corporations, and governments are doing into better-coordinated, collective efforts and reshape the trajectory of our global society.

A Shift in Perspective

✦ This is a book about perspective—the perspective of seeing our planet from space, the perspective of working on development projects on the ground, and the orbital perspective, which synthesizes them both and is the focus of this book.

There usually are two ways to define the word *perspective*. It may refer to the rendering or interpretation of a three-dimensional object on a two-dimensional plane, or it may refer to an attitude toward something, a point of view. Both of these definitions come into play in our discussion of the orbital perspective. Historically, for the most part, our perspective has been two-dimensional. Although we know that the world is not flat, a true perspective on a three-dimensional, interrelated reality is usually beyond our immediate awareness. Nonetheless, the real world is not two-dimensional, and to solve the problems facing our global society we need to start living in the real world. Two-dimensional thinking focuses on the next quarterly report or election campaign. Three-dimensional thinking, the orbital perspective, brings to the forefront the long-term and global effects of every decision. Two-dimensional thinking prioritizes capturing market share, whereas three-dimensional thinking looks to expand the market itself and does not particularly care who's got the biggest piece of the pie.

In my development work, I've seen how the two-dimensional, short-term approach can sometimes cause even the most well-intentioned efforts to go astray. Take, for instance, the situations

encountered by my friend Samantha Snabes, who we will hear
more about later. Samantha visited the Mugonero Hospital in western Rwanda, where in 2006 I had helped install solar panels with a
combined Manna Energy Foundation / Engineers Without Borders
team. While she was there, Samantha encountered an area behind
the hospital where a mountain of medical equipment, donated
by various organizations, was just sitting out in the elements.
These pieces of equipment had either broken because they weren't
designed for the conditions of what is effectively a field hospital,
or simply had not matched the needs of the hospital in the first
place. This senseless waste left a big impression on Samantha. "I
see this image in my head almost every night before I go to bed,"
she said. The director of the hospital told her that nongovernment
organizations had come with this equipment, "and when it doesn't
work, we can't fix it, and this is just where we put it until we can
figure out what to do with it."

Unfortunately, this is not an isolated incident; it is a situation
that repeats itself over and over again. At its core, the problem is
a lack of information. Well-meaning people and organizations do
not have all the pieces of the puzzle and often make recommendations that are not sustainable. They are operating with a two-
dimensional perspective. Without democratized access to information, individuals and groups in developing countries are forced
to depend on external as opposed to locally inspired solutions.

Samantha saw a similar disconnect at the L'Esperance orphanage, a short walk from the hospital, and in this case it was particularly troubling because the problem related to agriculture.
Subsistence farmers frequently make critical decisions about what
type of crop to plant based on limited or incorrect information,
and this had a specific impact at L'Esperance. Focused on creating
a financially sustainable income for the orphanage through dried
tropical fruit exports, and using a bit of agricultural experience
from his home in Guatemala, the director had borrowed a plan

from someone a few hills over and had invested the orphanage's limited resources and its workers in planting pineapples and other fruits. There was little consideration of factors such as soil type, rainfall needs, or soil acidity. "They literally gained access to a variety of seeds that looked like they might work, so they went ahead and planted them," Samantha said.

Unfortunately, they had planted the wrong type of pineapple for that area of Rwanda, and the crop would not produce sufficient fruit to meet the needs and goals of the orphanage. This is a serious concern for people existing on thin margins. While working in Nicaragua, Samantha observed a number of other situations in which impoverished people lost what little they had by planting the wrong crop, exposing impoverished families to tragedy after they used up their extremely limited resources in the hope of enabling subsistence farming. People's livelihoods—and at times lives— were being lost due to what boils down to a lack of information.

What Samantha saw in Rwanda, coupled with her experience in Central and South America, solidified her path. "I began to think that the best contribution I could make was to support decision-makers at the local level in developing countries by enabling solutions and technologies where people could leverage big data to make informed decisions and develop appropriate solutions," she said. So she set her sights on improving local food production.

A SHIFTING PERSPECTIVE

We all have moments like this in our lives, where something shifts, clicks into place. For me it was in June 2008, when I clamped my feet to the end of the robotic Canadarm2 on the International Space Station (ISS) and was flown through a maneuver that we called the Windshield Wiper, which took me in a long arc above the space station and back. As I approached the top of this arc, it was as if time stood still, and I was flooded with both emotion and awareness. But as I looked down at the Earth—this stunning, fragile oasis, this island that has been given to us, and that has protected all life from the harshness of space—a sadness came over me, and I was hit in the gut with an undeniable, sobering contradiction. In spite of the overwhelming beauty of this scene, serious inequity exists on the apparent paradise we have been given. I couldn't help thinking of the nearly one billion people who don't have clean water to drink, the countless number who go to bed hungry every night, the social injustice, conflicts, and poverty that remain pervasive across the planet.

Seeing Earth from this vantage point gave me a unique perspective—something I've come to call the orbital perspective. Part of this is the realization that we are all traveling together on the planet and that if we all looked at the world from that perspective we would see that nothing is impossible. It is possible to have peace in the world, to end wars, violence, and terrorism, but this can be done only once poverty is wiped from the face of the earth. And the good news is that it is also possible to lift the entire population of the world out of destitute poverty.

I don't claim to have a magic solution to all of our problems. Poverty and conflict are very complex issues, with numerous complicating factors, and I don't want to oversimplify these problems. But if we can address these issues correctly, if we can figure out ways to work together, it can propel us on a path to a more peaceful, safe, and prosperous world. One step in that unified direction is to put a greater emphasis on doing things that take a long-term focus and are financially sustainable, as well as things that make use of the power of the engines of enterprise.

I returned to Earth after that first space mission with a call to action. I could no longer accept the status quo on our planet. We have within our grasp the resources and technology necessary to solve many, if not all, of the problems facing our planet—and yet we don't.

I continued to ponder this during my second journey to space, which began in April 2011. I spent half of 2011 living and working aboard the International Space Station, and I spent most of any available free time I had with my face plastered to a window, gazing back at Earth. As I watched our beautiful planet, I wondered what the world would be like in the next fifty years, and I pondered a question that gnawed at me constantly: If we have the resources and the technology to solve the challenges we face, why do they still persist?

Up there, hovering above Earth with my orbital perspective, I came to believe that the problem lies primarily in our inability to collaborate effectively on a global scale. There are millions of organizations around the world working to improve life on Earth, but for the most part these organizations are not engaged in a unified, coordinated effort. There is a great deal of duplication of effort, loss of efficiency, and unfortunately, in many cases, destructive competition that does not lead to better products or services but only to substandard products and services, rising prices, and other detriments.

We already possess all the technology we need to enable truly consistent, world-changing global collaboration. Our real challenge lies in demonstrating how vital and valuable such collaboration is, despite the real and perceived risks. Open collaborations make solutions more effective and efficient through the pooling of resources and information. Working together multiplies cost-effectiveness while reducing duplication of effort. It is the only real way to enable economies of scale that can scale innovative solutions. Perhaps most importantly, open collaboration encourages greater accountability, which in turn fosters trust. We are all in this together, so we should share common goals, and the only way we are going to overcome the challenges facing our world is by working together.

An effective collaboration mechanism will pair critical challenges with vital solutions, bringing together unique pieces of the puzzle and enabling us all to learn from each other's successes and failures, making organizations' technologies and approaches considerably more effective than they would be otherwise. Since multiple organizations are looking to develop tools to enable collaboration, it is critical to unify those efforts.

As I looked back at our Earth from the orbital perspective, I saw a world where natural and human-defined boundaries shrank. I saw a world becoming more and more interconnected and collaborative—a world where the exponential increase in technology is making the "impossible" possible on a daily basis. Thinking about the next fifty years, I imagined a world where people and organizations set aside their differences and their destructive competitive inclinations—such as striving to maximize economic growth at all cost, or pillaging society for the personal gain of a few—and instead work together toward common goals. After all, we are all riding through the universe together on this spaceship we call Earth. We are all interconnected, we are all in this together, and we are all family.

I imagined a world where open, transparent collaborations become the engines that fuel tremendous economic growth and help us obliterate many of the problems facing our planet. I imagined that individuals and organizations that currently engage in destructive competition, secretive dealings, or corruption would begin to see themselves being left behind and would be compelled to adapt, evolve, and take on a much more effective collaborative focus in order to keep up with the economic growth that collaboration would bring. I imagined a world where we are all unified in the belief that by working together we can accomplish anything.

For almost all of human history, the vast majority of people believed that it was impossible to fly to the moon—simply because it had never been done before. Human ingenuity and human determination proved that it *was* possible. Today, the majority of people believe that it is impossible to solve many of the world's problems—that it is impossible, for instance, to lift the entire global population out of poverty. But if we adopt the same collaborative mind-set and practices that got us to the moon and back, and that built the International Space Station, we can alleviate poverty—and do much more.

NOTHING IS IMPOSSIBLE

When I visited L'Esperance in 2006 there was great need for infrastructure. The only available power was a small amount of electricity provided by a couple of solar panels. To get water, the children had to hike down and then back up a steep, muddy embankment and fill jerricans from a cistern filled with contaminated groundwater. When we tested the water supply, it had basically the same bacteria count as raw sewage. A nongovernment organization had come to the orphanage a few years earlier and installed a water pump and piping so the children wouldn't have to make the hazardous trek to get water. Unfortunately, the pump failed a few months after the NGO left, and the orphanage never had the funds to get it fixed. Another failure caused by well-meaning but short-sighted thinking.

The children's living conditions have improved greatly since that time, thanks to organizations such as Engineers Without Borders, the Manna Energy Foundation spinoff Manna Energy Ltd., and others. Each of the residential buildings has its own rainwater catchment system. Manna Energy Ltd. installed a large-capacity water purification system that utilizes gravel and sand filters and solar-powered ultraviolet purification, ensuring that all the children are drinking safe water. Engineers Without Borders also has worked with the community to construct open-air kitchens that use high-efficiency cookstoves, reducing wood use and the respiratory illnesses caused by indoor air pollution. These are just a few of the interventions that have improved the orphanage's ability to care for the children.

The first step toward effecting this kind of real change is to believe that real change is possible, and I believe it is possible because we already have accomplished some things that were once thought impossible. In the first part of this book, for instance, we will look at the historic U.S.–Russian collaboration in space, which in the context of the Cold War and its aftermath would have seemed impossible to many, and which demonstrated that collaboration is possible even in times of animosity and distrust. Although the focus here is primarily on the relationship between Russia and the United States, this collaboration eventually led to the fifteen-nation ISS partnership that embarked on the largest, most complex space construction project in history. Along the way, we will examine the kinds of issues that those early cosmonauts and astronauts faced in working together, and how they managed to overcome those issues, serving as a micro example of the principles that might inform our macro solutions to the world's problems.

The second part of the book explains more fully what I mean by the orbital perspective and the specific lessons to be learned from that perspective. Finally, in part 3, we will look at an incredible story of international collaboration, when NASA personnel and representatives from many distant nations and communities came together to develop solutions to a dire situation 2,300 feet below the surface, with the lives of thirty-three Chilean miners hanging in the balance. We also will discuss some of the specific issues we face in trying to create unified development efforts, where we could improve communication and sharing and eliminate destructive competition. Finally, we will look at some people and initiatives that are working to employ the orbital perspective to collect, share, and use information to resolve both local and global problems.

FACING A COMMON ENEMY

There are good people and organizations all over the world proving that if we all commit to working together we can live in a world without poverty, where no one dies from preventable and curable diseases, where everyone has access to clean water, where everyone's children can be educated. We are limited only by our imagination and our will to act.

In pulling back to the orbital perspective, it becomes clear that the first step toward solving our big challenges is to identify ourselves as a global community confronted with a common enemy. We must see ourselves as fellow co-laborers in a war against the many challenges we face in common. It is possible to replace the present wasteful, uncoordinated practices and futile go-it-alone mind-sets with collaborative practices that make the best use of available resources, that share information, and that effectively tap into the collective genius of more than seven billion creative minds.

In fact, this is not only possible, it is necessary, because we are fighting for our survival. The cause we are fighting for is more important than our own personal objectives or the objectives of our particular organization. We need to check our egos at the door and realize that overcoming our world's biggest challenges is more important than winning every battle, and that in order to address those challenges we're going to need everyone working and sacrificing together.

Our newfound technological tools of collaboration can unite us all in common purpose for the good of all. We can build bonds of camaraderie beyond those who are in our units or our organizations, or who wear the same uniform as we do. Having a shared purpose can enable camaraderie and the resulting collaboration that will be required to solve the problems facing our world. The International Space Station exists because nations and people put aside their differences to accomplish a common goal. The Chilean mine rescue was successful because of the esprit de corps that developed among everyone involved. We need to expand the circle of people we care about and replace the ripples of suffering with ripples of hope, in the words of the late Senator Robert F. Kennedy: "Each time a man stands up for an ideal, or acts to improve the lot of others, or strikes out against injustice, he sends forth a tiny ripple of hope, and crossing each other from a million different centers of energy and daring, those ripples build a current which can sweep down the mightiest walls of oppression and resistance."[1]

Applying an orbital perspective to global problem solving starts with identifying the fundamental causes that we are willing to work and sacrifice for, seeking out individuals and organizations that share our purpose, and then working together with them for the cause. It means supporting others when they do great work, even if it means their organization might have a better chance of getting a grant that your organization is also seeking. It means having the humility and courage to identify those areas where you or your organization are lacking, and seeking out ways to partner with others by offering the value that you do have in other areas. The orbital perspective requires taking a long-term view, which at times requires sacrificing short-term progress for the achievement of overarching goals. Above all else, we must realize that we need each other. The days are long gone where a single nongovernment organization—or a single nation, for that matter—can effect the change that's required to solve the problems facing our world.

We don't have to be in orbit to have the orbital perspective. We just need to apply that perspective to our work here on this planet. And if we commit to working together, we won't have to accept a status quo far beneath the potential of this beautiful world.

PART I

LOOKING SKYWARD

Humanity's Home in the Heavens

✦ On July 17, 1975, at 7:19 p.m. GMT, Soviet cosmonaut Alexey Leonov and American astronaut Tom Stafford reached across the hatches of their docked Soyuz and Apollo spacecraft and shook hands 140 miles above Earth. The event, which represented the end of a long, expensive space race and the beginning of a movement toward the peaceful exploration of space, was the end result of an agreement forged in May 1972, when President Richard Nixon and Soviet Premier Alexei Kosygin formalized a commitment to making a peaceful joint program of space exploration a reality. Speaking on the significance of this agreement, Soviet leader Leonid Brezhnev noted, "The Soviet and American spacemen will go up into outer space for the first major joint scientific experiment in the history of mankind. They know that from outer space our planet looks even more beautiful. It is big enough for us to live peacefully on it, but it is too small to be threatened by nuclear war."[1]

The Apollo–Soyuz mission was heralded as a breakthrough in Cold War diplomacy, but the collaboration was short-lived, and the end of the mission marked the end of the two countries' real cooperation in human spaceflight for nearly two decades. According to George Abbey, former head of the Johnson Space Center, the Soviets wanted to continue working with the Americans on joint missions after the Apollo–Soyuz mission, but the Americans did not wish to continue. Instead, the Americans saw their own space shuttle on the horizon, with its revolutionary promise of relatively safe, inexpensive access to space and a flight rate of fifty to sixty missions per year. It was envisioned that the shuttle would herald a new era of U.S. space exploration, including enabling the construction of a massive space station. With all these things on the horizon, the United States didn't see a compelling reason to continue to partner with the Soviets.

Over the next two decades, the Soviets continued their pioneering work in space station launch and design, which had begun with the launch of the first space station in history, *Salyut 1*, in 1971. A couple of years after the Apollo–Soyuz docking, in 1977, the Soviets created the second generation Salyut station, followed in 1986 by the construction of Space Station *Mir*, the name of which means "peace" or "world." Meanwhile, the United States was building its space shuttle and pursuing its goal of building Space Station *Freedom*.

Unfortunately, the space shuttle would never live up to its promise of being inexpensive, safe, or easy to operate at a high frequency, and because of the shuttle's shortfalls, as well as a change in political will and funding, the dream of constructing a massive, highly capable U.S. space station languished. Since the early 1980s, roughly $11.4 billion had been spent,[2] and the station had been redesigned several times, but by 1992 no hardware had been delivered to space, and congressional support was drying up. Space Station *Freedom* was most likely going to die before a single component had been launched, and even if it didn't get canceled, it would be over budget and way behind schedule. On the Russian side, *Mir*, which was scheduled to be superseded by *Mir-2*, was also in trouble. With the collapse of the Soviet Union and subsequent financial problems, it was apparent that the Russians would not be able to afford to launch and assemble *Mir-2*.

Thus, by the early nineties the geopolitical and space program planets aligned and the time was ripe to readdress a Russian–U.S. partnership in space exploration. The rudiments of the planned

Space Station *Freedom* and the planned Space Station *Mir-2* could be repurposed into an international program. The Americans would gain experience docking shuttles to a large station, and with the Russians, would develop the docking system that would eventually be used on the International Space Station (ISS).

The plan, which became known as the Shuttle–*Mir* program, immediately leveraged both the U.S. and Russian space programs. All of a sudden, we would have two spacecraft capable of carrying humans into space, and each space program would bring unique solutions to different pieces of the puzzle, adding value to the partnership. The Americans had little experience in operating a space station; the Russians had vast experience. The Russians also knew how to build modules cheaply, and billions of dollars could potentially be saved by merging the Russian and American space station programs. The American space program, on the other hand, was much better funded, and the space shuttle could provide badly needed resupply to the aging *Mir* and serve as the long-term workhorse for construction of the ISS. Two big programs that were headed toward the cliff could be salvaged.

COOPERATION AND TURMOIL

In June 1992, President George H. W. Bush and Russian President Boris Yeltsin signed an agreement, which evolved under the Clinton administration, to undertake the Shuttle–*Mir* program, the first phase in a long-term plan for cooperation in space. Plans to build the U.S.-led Space Station *Freedom* would be abandoned and Russia would join a new partnership to design and build what eventually became the International Space Station. This program, in the eyes of politicians in both countries, solved myriad problems. Russia saw its partnership with the United States, Europe, Canada, and Japan as a way to achieve acceptance by Western nations while keeping its deteriorating *Mir* program going. Perhaps the most important benefit, in the U.S. view, was that pumping money into the Russian space program would discourage Russian rocket scientists and missile technology from being exported to countries with a desire to do harm to the United States and its allies. This was a particular concern, given the recent collapse of the Soviet Union and the uncertain future in that country.

A number of technical meetings followed the Shuttle–*Mir* agreement. In July 1993, for instance, Yuri Semenov, chief of Energia, the now-commercial former Soviet Design Bureau, held a symposium in the United States to discuss *Mir* and to attract interest in and business to the Russian space station. Semenov gave approval for the top Energia people associated with *Mir* to attend and, for the first time, to speak openly about the space station and its capabilities. It was an extraordinary, historic, and successful symposium—a rich technical exposé of this orbiting facility that had been shrouded in secrecy.

Following the symposium, the Russians and Americans began a series of meetings in Crystal City, Virginia. These meetings occurred in a nondescript, nameless conference room in a building that both the Russian and American participants referred to as oden, oden, oden, because *ah-dyin* (один) is the Russian word for "one" and the building's street number was 111. The goal of these meetings was to explore ways the United States and Russia could collaborate in space and would prove to be a critical turning point in the two nations' destiny in space. These initial technical meetings both indicated and fostered a real desire to work together, which is probably best illustrated by the lengths to which people from both countries were willing go to ensure the project kept moving forward.

For instance, although Russia was in a state of turmoil and the country was seemingly falling apart, George Abbey, NASA Administrator Daniel Goldin, and other NASA personnel traveled to Moscow in October 1993 to meet with the Russian Federal Space Agency, Roscosmos, and negotiate Russia's entrance into the ISS program. The group arrived in the middle of a ten-day conflict in which, according to government estimates, 187 people were killed and 437 were wounded; estimates from nongovernment sources put the death toll as high as two thousand. President Yeltsin, facing impeachment by many members of the Russian Duma, called on the military to end the rebellion. Despite the turmoil, meetings were conducted at Roscosmos headquarters in the northern part of Moscow, just outside the Garden Ring.

On October 3, the Russian and American delegations gathered around a large, round table while gunfire echoed outside. Goldin recalled watching TV and thinking it was like being in a movie. "You could see people bashing in buses, throwing them over. You could hear machine guns firing."[3] That evening, armed pro-parliament demonstrators advanced on the Ostankino television center, not far from the Penta hotel where the NASA contingent was staying. As the demonstrators approached the TV complex, military units met them and a fierce battle ensued. Part of the TV center was significantly damaged, television stations went off the air, and sixty-two people were killed.

Goldin called Washington for guidance on whether the delegation should leave the city. Soon a call came back: "The President of the United States would like to show support for democracy in Russia. If it's safe, please stay." Goldin called the group together to take a vote. Every single person voted to stay.

By sunrise the next day the Russian army had encircled the parliament building, and a few hours later army tanks began to shell the Russian White House. Meanwhile, the NASA contingent was getting ready to head from the hotel to resume negotiations at Roscosmos headquarters. Abbey recalled seeing tanks firing into the White House on television and then, a few seconds later, hearing the blasts and seeing the smoke rising outside.

Even in the presence of this much instability and uncertainty about the future of their country, the Russians were still willing to talk about cooperation in space. Likewise, the effort was so important to the Americans that they were willing to risk life and limb to push the partnership forward. Space rose above the fray. As Goldin put it, "That day of revolution, we negotiated the Russian entry into the International Space Station." There was great motivation on both sides to move things forward because, in reality, both sides needed each other.

COLLABORATION IN SPACE

The Shuttle–*Mir* program began in earnest on February 3, 1994, with the launch of Space Shuttle Mission STS-60. On the middeck of Space Shuttle *Discovery*, Russian cosmonaut Sergei Krikalev was making his third trip to space. Krikalev would go on to fly three more space missions, including two ISS missions, and ended up spending more time in space than anyone in history—an astonishing 804 days.

A year and three days later, on February 6, 1995, during the STS-63 mission, Shuttle Commander Jim Wetherbee maneuvered *Discovery* to within 36 feet of Space Station *Mir*. On board *Mir* were *Mir*-17 commander Aleksandr Viktorenko and cosmonauts Yelena Kondakova and Valeri Polyakov. There was to be no docking on this mission, just a close approach and partial flight around the station. The mission was a dress rehearsal of sorts for the first docking to the station.

As Wetherbee approached *Mir* he radioed, "As we are bringing our spaceships closer together, we are bringing our nations closer together. The next time we approach, we will shake your hand and together we will lead our world into the next millennium." Viktorenko responded, "We are one! We are human!"

For Mike Foale, who was on board *Discovery*, the realization of the importance of the mission actually came later:

> We all thought this *Mir* thing was kind of a jaunt—it was just an add-on [to the mission], albeit exciting. But it was when MCC-Houston [Mission Control Center, Houston] sent up a crappy picture made from a TV picture downloaded from *Mir* to MCC-

Moscow—a really bad black-and-white image of us on the shuttle, coming up to the *Mir*—that it suddenly dawned on us that we had done something really important. It was that view of us from the Russians' point of view that brought out a key issue in collaboration, and that's looking at the world through the other person's eyes.

The first shuttle docking to *Mir* came in June 1995, during STS-71, when Shuttle Commander Robert "Hoot" Gibson docked Space Shuttle *Atlantis* to the station. This was the first time that the U.S. space shuttle, which was designed to construct and dock to an American space station, had ever docked to anything. The mission delivered Russian cosmonauts Anatoly Solovyev and Nikolai Budarin to *Mir* and gave American astronaut Norm Thagard a ride home following his historic first U.S. mission on board *Mir*. He had launched to the space station nearly four months earlier aboard a Russian Soyuz spacecraft.

BUILDING A FOUNDATION OF TRUST

The Shuttle–*Mir* program continued until 1998. In all, seven U.S. astronauts spent nearly one thousand days on the orbiting space station. At times the relationship was strained by various emergencies and crises. In 1997, for instance, a fire broke out on board, threatening the life of the crew, including American astronaut Jerry Linenger. Later that year, the unmanned Russian cargo ship *Progress* collided with the station. The collision breached *Mir*'s hull and sent the crew, including American astronaut Mike Foale, scrambling to cut off the damaged part of the station and isolate the subsequent depressurization as the space station's air leaked out into the vacuum of space. Incredibly, despite the severity of the situation, the crew was able to isolate the leaking section of the station, but the aftermath of the collision left the orbiting complex tumbling and without power for many hours.

Such life-threatening events put a serious strain on the U.S.–

Russian relationship, but they also helped forge a degree of trust. In the hours after the cargo ship collision, for instance, the crew had no choice but to collaborate to save their lives, and after the initial emergency was over, the United States and Russia improved their national collaboration out of necessity, to save the Shuttle–*Mir* program. The Russians became more willing to share technical data about their space operations, and the Americans came to prove they were in it for the long haul. The Americans demonstrated that they believed the partnership was valuable enough to continue flying American astronauts to the station, even though the continued U.S. presence on board *Mir* was a very controversial decision fraught with political bickering.

The decision of all parties to stay the course proved to be a critical component in building the trust that was to become the foundation of the International Space Station program. The lessons learned during the Shuttle–*Mir* program enabled the fifteen nations of the International Space Station partnership—which includes Canada, the nations of the European Space Agency, Japan, Russia, and the United States—to embark on the largest, most daring peacetime international collaboration in history.

The first component of the ISS was launched in 1998, and the station has been continuously inhabited since November 2000, surpassing the previous record of nearly ten years' continuous human presence in low Earth orbit—held by *Mir*, of course.

However, the ISS program has also had its share of challenges to overcome to keep the partnership together, some of which we will detail in the next chapter. No challenge was more dire and critical, though, than the aftermath of the morning of February 1, 2003—the day Space Shuttle *Columbia* disintegrated on reentry, killing the crew of Rick Husband, William McCool, David Brown, Kalpana Chawla, Michael Anderson, Laurel Clark, and Ilan Ramon. In the wake of the disaster, the shuttle fleet was grounded for more than two years while new safety measures were incorporated into the design.

During this period, the Russians picked up the slack, transporting crew members to and from the ISS. But as the grounding of the shuttles dragged on, there was much concern among the ISS partners that the United States would be unable to fulfill its commitments to the construction of the space station. In June 2005, newly appointed NASA Administrator Mike Griffin faced this concern head-on at a meeting of ISS partners in Paris. Essentially, Griffin's counterparts from the partner space agencies were saying, "If you don't complete the ISS, we will never work with you again, because we will have all lost our jobs." The international partners were very concerned, and they were very aware that the decision to finish the space station was seriously in doubt.

Griffin knew that the United States had to live up to its commitments to the station, which included launching the Japanese and European modules. Failure to do so would be disastrous to the international partnership, which he considered the most important part of the ISS effort. But doing so would not be easy. According to Griffin, "What the OMB [Office of Management and Budget] told me when I took over NASA was that I only had enough money for fifteen more shuttle flights, post-*Columbia*. . . . It was very clear to me that it was essential for us to meet our commitments and get those modules launched. But it was also clear that I wasn't going to have enough shuttle flights to get everything done, and there was nothing I could do about that." At the time there were nearly thirty shuttle flights on the books to complete the construction of the space station.

In the end, twenty-two shuttle flights were flown to complete the construction of the ISS. To accomplish these flights, Griffin needed to be creative, within the constraints of the NASA budget. "I took money from science and other places, and I got yelled at for that—but I did it." And his creativity paid off. "Because I made the decision quickly, we stuck to it, and the actions that followed were obviously in support of the decision, it engendered a lot of trust on the part of our partners, for the United States and for NASA."

Until that decision was made, however, it was not clear that the United States was going to commit to completing the ISS.

Following completion of the ISS construction in the summer of 2011, the space shuttle program was retired. I had the honor of closing the hatch on that last shuttle flight and ringing the ship's bell on board the space station as Space Shuttle *Atlantis* and the crew of STS-135 undocked and departed from the ISS for the last time. This was a very emotional experience for all of us. Closing that hatch signified the closing of a long chapter of the U.S. space program, a chapter filled with hard-won successes. But it also signified the fulfillment of a commitment and the opening of a new chapter in space exploration, beyond low Earth orbit.

This new era of cooperation is the true fulfillment of a goal articulated nearly fifty years before that final shuttle flight to the ISS. In September 1962, President Kennedy famously stated:

> For the eyes of the world now look into space, to the moon and to the planets beyond, and we have vowed that we shall not see it governed by a hostile flag of conquest, but by a banner of freedom and peace. We have vowed that we shall not see space filled with weapons of mass destruction, but with instruments of knowledge and understanding. . . . There is no strife, no prejudice, no national conflict in outer space as yet. Its hazards are hostile to us all. Its conquest deserves the best of all mankind, and its opportunity for peaceful cooperation may never come again.[4]

The story of this international cooperation is largely a story of technical, political, personal, and other challenges, and of the trust earned among all the partners in the ISS program, which will be detailed in the next chapter. Out of those challenges, moreover, we can discern a number of principles that may be used to guide cooperation both in space and on Earth, which we'll get into in chapter 3. These principles, and the motivation to adhere to them, are inspired by the orbital perspective.

CHAPTER TWO

Space, the Shared Frontier

✦ Early one morning in March 1994, thirty-five-year-old NASA flight surgeon Mike Barratt was riding in the backseat of a Soviet-era Volga with a cracked windshield. He was traveling from Moscow to meet with his counterpart at Roscosmos, Russian flight doctor Igor Shekhovtsov. Barratt, who in 2000 would be selected into the eighteenth class of astronauts, along with me and fifteen others, was assigned as the flight surgeon for astronaut Norm Thagard's historic first U.S. mission aboard *Mir*. As flight surgeon, Barratt was responsible for coordinating all medical aspects of the mission and ensuring that Thagard met preflight medical requirements and stayed healthy before and during the mission. Barratt also was responsible for all the medical experiments that would be performed by and on Thagard. While in space, astronauts serve both as lab technicians, conducting and monitoring experiments, and as lab rats, the subject of experiments. The micro-g environment provides a unique opportunity to further our understanding of the human body.

The Volga made its way along dark, winding back roads with no streetlights, past dachas and small industrial towns. Mounds of exhaust-darkened snow lined the roads. Barratt's driver was one of the typically friendly, talkative, and extremely capable men employed to shuttle the Americans between Moscow, the Gagarin Research and Testing Cosmonaut Training Center in Star City (about 50 kilometers northeast of Moscow), and Russian mission control, in the industrial city of Korolyov. Road conditions in Russia in the early days of the U.S.–Russia space partnership were like something out of the movie *Mad Max,* and the Star City drivers were so aggressive that they were affectionately called "double-O" drivers—because, just like James Bond, each had a license to kill.

After a little over an hour on the road, Barratt and his driver passed a Soviet-era fighter jet perched as a monument on the side of the road, with the inscription "слава советским покорителям неба! [Glory to the Soviet Conquerors of the Sky!]." Nearby was a large sign that read звездный (shorthand for Star City), where cosmonauts have lived and trained since the 1960s. The site once had been top secret—reportedly known as Closed Military Townlet no. 1, among other names—and it was still heavily guarded. Once inside the gates, the battle-hardened Volga drove down a beautiful tree-lined road steeped in history. Every cosmonaut since Yuri Gagarin, the first human to orbit Earth, had begun the journey to space by traveling along this road. It cut through Soviet-era apartment complexes for cosmonauts, staff, and their families, past walking trails, parks, and ponds to the nearby Chkalovsky Airport (чкаловский). From there, cosmonauts traveled to the Baikonur Cosmodrome, in what is now Kazakhstan, for launch. Now, for the first time in history, an American astronaut would be joining them.

After passing through a second military checkpoint, Barratt arrived at Building 3, the medical building inside the main functional area of the training center, and one of the better-maintained buildings in the compound. In most other buildings, it was not uncommon to find missing floorboards, lights and bathrooms that didn't work, and cats running through the buildings—to take care of the mice, of course.

Inside Building 3 Barratt met with his counterpart, Shekhovtsov, a bear of a man. Not long after the meeting began, Barratt and Shekhovtsov were discussing medical protocol, comparing the differences between Russian and American protocols for space crew members. Suddenly, a door flew open and a man dressed in the uniform of Russian Air Force colonel burst in and said sternly, "It is forbidden for you to discuss that." He immediately left and closed the door behind him, without any further discussion.

Barratt sat in stunned silence. It was clear that the man had been listening to the two doctors' conversation. But his Russian colleague seemed unfazed, as though nothing had happened, and as the joint space operations progressed, it became apparent that such surveillance was not uncommon. Tom Marshburn, another flight surgeon-turned-astronaut, who was also in Russia during the early days, and astronaut Mike Foale, who spent 145 days on board *Mir*, both noted it as well. It was also apparent that not only did the Russians not think this was unusual, they didn't even try to hide it. In fact, Foale knew that the Russian management was eavesdropping on NASA personnel because the Russian general in charge of Star City admitted as much to him.

This was somewhat typical during these early days of the joint space program, when there was still a general mistrust on both sides. For many people in both countries, the Cold War had never ended, and they were working with their counterparts from the other side only because they had been ordered to do so by their respective political leadership. As Barratt put it, "The Cold War didn't die all at once. Old habits die hard."

Foale himself was greeted with suspicion when he first started training with the Russians. Shortly after he arrived at Star City, one of the top Russian managers and his deputy sternly instructed Foale that he was not to leave the base, was not allowed to go anywhere. Moreover, if he was ever invited anywhere, Foale was to inform the manager. Foale acknowledged that this was a very aggressive tactic, but he also had an interesting view of the eaves-dropping and surveillance: "Collaboration works when the guy can see the other position. The generals over in Star City were actu-ally rather good at seeing our point of view, partly because they were bugging us. In a really weird and bizarre way, the Russian management had a better understanding of our issues than our management did of them, because the Russian management had direct access to apartment bugging."

TURNING POINT

The Americans heading to Russia were familiar with security measures—NASA has its secure facilities as well. But this level of overt surveillance was new to them. Likewise, the aging infrastructure and vehicles, the double-O drivers, and other cultural differences and misconceptions represented hurdles to overcome, and came as something of a shock to many in the early days of the partnership.

A major difficulty was each side's attitude toward—and at times, lack of respect for—the other. Some Americans went to Russia in the early days believing that the United States was superior to Russia in almost every way. In their opinion, the United States had won the Cold War, just as it had won the space race by landing on the moon. The United States also had the beautiful, highly capable space shuttle. In their view, the United States was only acting as a political and financial tool to prop up the failing Russian space program and to keep Russian rocket scientists from selling technology to hostile nations. In general, they felt the Russians—and the Russian space program in particular—were a little on the primitive side. What could the Americans really learn from the Russians? The state of the Star City facilities and some other parts of the country at that time certainly seemed to back this opinion.

Unfortunately, the people who were trying to "fix" the Russian program mistook a *different* way of doing things for an *inferior* way of doing things. They also overlooked the amazing accomplishments that the Russians had achieved during the previous five decades of their space program. At times this mind-set led some in the U.S. camp to deal with the Russians in a very condescending way, which understandably was not received very well. As Barratt put it, "Some of our people who did the worst over there went over there to help them see the backwardness of their ways and to reform their program. And at very high levels that just did not work."

On the other hand, many Russians were justifiably proud of the Russian accomplishments. They had, after all, launched the first human into orbit. They also had a history of long-duration space-flight on a series of Russian space stations, whereas the United States had very limited long-duration spaceflight experience. More-over, some Russians felt that their experience, earned over decades of literal blood, sweat, and tears, was part of the Russian national treasure and shouldn't just be sold away to the Americans. So, some of the Russians were likewise condescending toward their American counterparts. Foale mentioned, for instance, "During the first few technical lectures, there was dripping condescension from some of the instructors, and I was just stunned. They would say, 'You Americans, you do spacecraft all wrong.'"

On the whole, however, whereas many on the American side saw the partnership as a political gesture directed by the White House, or as the United States saving the Russian space program, the consensus among many of the Russians seemed to be that the two parties were saving *each other*. In practice, this meant that the Russians were willing to help the Americans in cases where they felt Russia had more experience or know-how, and vice versa, but both the Americans and the Russians were at times unwilling to hear or accept such assistance.

Jeffrey Manber, who at the time was the head of RSC Energia-USA, recalled a meeting with Leonid Gorshkov, one of the design-ers of *Mir*, in which these impasses seemed to take a critical turn for the better. The Russians, as I've mentioned, had vast experience in the construction and operation of space stations. Before *Mir*, the country had a number of others, which had grown in complexity and had evolved over the years. The Russians knew what they were doing as far as construction, and they knew a great deal about the long-term habitation of space stations. The Americans, in com-parison, had considerably less experience with space stations, with only *Skylab* to their credit and only three manned missions to the station, conducted in 1973 and 1974.

Gorshkov and his colleagues were therefore in a position to spot critical fundamental flaws in the design of Space Station *Freedom*. At this meeting, Gorshkov took out a piece of paper and sketched *Freedom*.

"Friends, look," he said. "Here's the problem with *Freedom*." And he proceeded to outline the Russians' concerns. He then made another drawing and said, "But friends, we can do it this way. If you did this and this and move this truss here, this is better. And we could help you here, here, and here, based on what we've learned from *Mir*." He slid the second sketch across the table.

"And the Americans," Manber said, "look at it and go, 'That's interesting.'"

In Manber's view, the Russians understood what was at stake. They seemed to acknowledge that both sides in the space community needed each other. And at this turning point the Americans started to realize the value of the Russians' space station experience. As it turns out, without this partnership, the ISS would not have come to be, nor would the United States have been able to keep the ISS program going after the *Columbia* accident. The two countries were in fact saving *each other's* programs.

COMMUNICATION BREAKDOWN

A negative attitude or perceived lack of respect for the other side is the result of one kind of misunderstanding or miscommunication, but lack of a literal common language was understandably a big issue, too. The vast majority of the Americans who traveled over in the early days did not speak Russian, and according to Russian Elena Maroko, chief interpreter in the Moscow Liaison Office, they were meeting with Russians who had not had much opportunity to learn or practice English. The average Russian working with the Americans in the early days of the partnership did speak English much better than the average American spoke Russian, and many Russians were more familiar with American literature than the Americans were, but the Russians were not allowed to travel abroad because the space program was considered defense related. "The likelihood of those people going abroad and practicing so their language skills could improve was basically nonexistent," said Maroko. "So language was probably the biggest barrier that people had to overcome"

In some cases, the language barrier nearly led to some serious consequences. For instance, Foale's first technical lecture, on how to control and maneuver the Soyuz spacecraft in orbit, was taught in Russian, without any interpreters. Unfortunately, there is a common confusion that most non-native-Russian-speaking students (myself included) get into when they learn about monitoring spacecraft control: there is a Russian word that sounds like *control* but actually means "monitor," whereas the Russian word for "control" sounds nothing like *control* in English. The resulting confusion is understandable, but it was causing Foale's instructor some frustration.

At one point the instructor turned to another Russian in the room and said, *"Kakoi durak!* [What a fool!]," along with some other choice words.

Foale, very angry, responded, *"Ya ne durak!* [I'm not a fool!]," startling the instructor.

One of Foale's Russian colleagues, offended on his behalf, yelled back in Russian, "How can you say that? He's perfectly ready to hear your lesson. You don't understand what Mike is trying to explain to you."

Foale's later assessment of the situation—that the instructor wasn't looking at his lesson from the point of view of a student who doesn't speak his language—seems reasonable, but three years earlier the Americans had subjected the Russians to a similar oversight, in an even bigger way, during the historic signing of the first contract between NASA and Energia. The contract was an agreement for NASA to use the Russian Soyuz spacecraft as an emergency escape "lifeboat" for the planned Space Station *Freedom* program, and the document was being hand-delivered by Jeffrey Manber.

Manber walked into the office of Energia chief Yuri Semenov and placed the contract on the very table where Energia had made the decision to launch *Sputnik* and to put the first human in space, and according to Manber, there was "general euphoria" in the room. Unfortunately, the initial euphoria quickly turned to disappointment when the Russians examined the document. It was written only in English. There was no translation. The Russians saw the lack of effort to translate this extremely important, even historic, document into their language as a major slight.

In retrospect, I also find this oversight shocking, but as the United States and Russia have learned to work together we have learned the importance of dual-language documents. In fact, all emergency procedures (and many other documents) used on board the ISS are written in English, with the corresponding Russian translation on the opposing page.

CULTURAL MISUNDERSTANDING

In addition to a lack of respect and trust, compounded by a lack of common language, one of the biggest issues in the early days of the Shuttle–*Mir* program was cultural misunderstanding. For instance, Barratt noted, "Culturally, the Russians are very different from us. They think very differently; they think very pragmatically. Their emotions are worn on their sleeves; they're not six layers deep in analysis." In the eyes of the Russians, on the other hand, the Americans sitting in negotiations and meetings were flat and emotionless, as though the Americans didn't really think what they were all trying to accomplish was important. The Americans considered this to be a solid, calculated negotiating tactic, but the Russians took it as apathy.

"Once we got a handle on that and we started negotiating much more aggressively, more like they would with one another, we started making progress," Barratt said. He recalled an instance when the Russian medical officials, who didn't trust the Americans' medical reports and tests, accused his partner of falsifying a sample, because the Russians believed the result should have been different.

"I got pretty mad," Barratt said, "and I started pounding my fist and I got red in the face. I yelled at this guy. And afterwards, it was like we were old friends, and we were toasting each other. I asked him, 'Why did I have to do that for you to understand?' and he replied, 'Well, we didn't think you cared.'"

Deep misunderstandings about the other side's culture and expectations appeared in other areas, too. For instance, Manber observed that, in the early days of the partnership, the Russians tended to respect age and seniority. Therefore, the Russians sent mostly senior Energia people, typically veterans of the Apollo–Soyuz program. The older, more experienced NASA people, on the other hand, didn't want to go to the Soviet Union/Russia. In Manber's estimation, "They didn't want to endure what many perceived at that point as poor living conditions, and they didn't even want it on their career stamp that they did business with the Soviet Union in ninety-one or Russia in ninety-two."

In response to this lack of volunteers in senior positions, NASA sent mostly young people to work with the Russians. For Manber, who was himself in the room because the Russians considered this partnership important enough to bring in an American with a deep understanding of the space community, it was difficult to witness this imbalance. "It was just so difficult to sit there, to see this thirty-five-year-old, fresh from Houston, with a brand-new passport, having never been overseas, trying to reach an agreement with Energia. It was frustrating for all of us, including the Russians."

Another critical kind of cultural misunderstanding is probably best illustrated by the story of a U.S. congressman who visited Russia shortly after the NASA–Energia contract was signed. Manber noted, "Americans like to begin the new year fresh. They like to go on business trips in the new year and get major business done. Well, that's the Russian holiday season. Their Christmas is the seventh of January."

The congressman arrived in Russia on January 8 and was upset that the Russians would not meet him. When Manber explained that this was Russian Christmas, the congressman replied, "Oh, they don't believe in that stuff. They're not religious."

The congressman thought the Russians were slighting him. The Russians, on the other hand, were proud that in this post-Soviet Russia they could celebrate religious freedom and a religious holiday, so they were staying at home to show their belief in the new Russia. They weren't going to come out to meet a congressman from another country during their Christmas.

"Would you go out to a congressman from another country on December 25?" Manber asked. "You would think it was rude of that person to come on December 25 and not know your calendar. My takeaway is that we have to take the time to understand the history, the motivation, and the needs of the other side."

EMPATHY IN COMMUNICATION

A final difficulty in the early days of the Shuttle–*Mir* project was simply the difficulty of gathering or relaying information. There was no Internet, for instance, which made it difficult to discover what any organization was doing or how it was structured, who headed it or how to contact them. One person who helped bridge this gap was Elena Maroko.

"I played the role of trying to establish those contacts, which proved to be extremely precious later on. It was very hard to understand the hierarchy. There were no organizational charts that were distributed; there were no documents that said who the department heads were or who reported to whom. So you had to draw your own charts. I'm still using some of these charts today."

These issues were perhaps unavoidable, given the technology and resources of the time, but there were instances where information conveyed to the other side was simply confusing or not useful. Take, for instance, the contract for the use of the Soyuz as a lifeboat for Space Station *Freedom*—the contract, you'll recall, that was written only in English. Not only was this massive document not translated, it also contained references to the U.S. Federal Acquisition Regulations, many of which either did not apply or were simply ludicrous.

Accordingly, two days after Manber delivered the contract, Energia's deputy chief designer, Alexander Derechin, called Manber into his office and asked, "What is the meaning of this clause that we must have American poultry, and what is the meaning of this clause that our bathrooms must be cleaned two times a day or that we have to print on both sides of the paper?"

At that time, U.S. companies were used to dealing with the requirements of FARS, but this was all new to the Russians, and no one at NASA had taken the time to explain to them what the regulations even were. As Manber noted, it's not difficult to imagine what it would be like for these men and women, already nervous about entering into an agreement with the United States, not only to receive such important documents in a foreign language but also to discover these strange, unexplained requirements. Clearly, this would not tend to create an atmosphere of trust and goodwill.

One FAR, in particular, stood out. It said, roughly, that the signee of the agreement would agree not to do business with Cuba, China, or the Soviet Union. Derechin said to Manber, "Jeffrey, I can promise you we won't do business with Cuba. I can pretty well promise you we won't do business with China. But Jeffrey, it's hard for me to promise we won't do business with ourselves."

Manber went back to NASA and told them the Russians wanted the FARS changed, to which his NASA contacts responded, "You can't change the FARS. They're the FARs!" The Russian mind-set, on the other hand, was that if the United States wanted to do business with them, they should just call up Congress and get the rule changed.

Eventually, Manber was able to bring both sides together and resolve the issues, but as he said, "The first really overarching conclusion that I came away with is you can't solve problems until you understand the other side. You cannot parachute into a particular part of the world or a particular industry and seek to solve a problem without putting yourself completely in the other person's situation."

In Manber's eyes, NASA had acted horribly in the beginning. Over the course of many years of working together, however, this early mistrust and misunderstanding has been replaced by a fully functioning, collaborative team that today works together as a finely oiled machine (most of the time). Everyone I have spoken with about working with the Russians in space exploration believes that the United States has learned a great deal from Russia and that Russia has learned a great deal from the United States— and that the entire international space partnership is much better off because of it.

But how did we go from those early days of mistrust and suspicion to building and operating the International Space Station? What was the secret ingredient that enabled a coalition of fifteen nations to work together systematically, in a fully integrated manner, to construct and operate the most complex structure ever built in space? Is there something we can learn from how the members of the ISS program learned to overcome differences and cultural misunderstandings in order to accomplish remarkable things together? And can we use these same techniques to reach agreements on things such as alleviating poverty, mitigating climate change, or achieving peaceful solutions to long-term conflicts? I believe that we can learn some lessons from this experience, and these lessons are the topic of chapter 3.

CHAPTER THREE

Lessons in Collaboration from the ISS Program

✦ In looking at the history and operation of the International Space Station partnership, certain lessons bubble to the top. For instance, we saw how an initial lack of respect or mutual understanding on both sides led to conflict or confusion in the early days. We also saw the importance of effective information sharing and communication.

Effective collaboration requires mutual understanding and a commitment on both sides to forming relationships and working toward the long-term success of the partnership. Partners who put in the necessary work and are in it for the long run develop real relationships and establish trust. As a result, not only do such partnerships survive when the inevitable bumps in the road occur, they usually are strengthened, and this enables the partners to achieve great things.

CULTURAL EXCHANGE

Collaboration begins with mutual understanding and respect. As we saw, one major problem in the early U.S.–Soviet/Russian collaboration was misunderstanding on both sides. Whether it was a language barrier, an unfamiliar way of doing things, or an underestimation of the other side's abilities, overcoming such misunderstandings required people on both sides to move outside their comfort zones and to really absorb another culture. There was a need on both sides to understand a style of communication that sometimes was very different, in addition to the language itself, and to respect others' limitations in these areas.

NASA, Roscosmos, and the other international partners all have taken extensive steps to incorporate the lessons learned during the Shuttle–*Mir* program. NASA established comprehensive language and cultural training programs, including immersive programs in which U.S. and other international astronauts live with a Russian family for a period of time. In my own immersion training, I learned a great deal about Russian language and culture by living for a month with a Russian family in an apartment in Moscow. Likewise, intensive English programs were initiated for Russian cosmonauts and trainers. The benefit of increasing the ability of people to communicate in formal and informal situations has been tremendous.

But moving beyond one's comfort zone to understand another culture could be as simple as trying different foods. Mike Barratt grew to love Russian food, but many of the engineers in the early program seemed unwilling to experiment, and even brought over suitcases full of things like tuna fish, Pop-Tarts, and Skippy peanut butter. Barratt thought this indicated a certain lack of resolve or seriousness on the part of the Americans, and he told his colleagues, "You guys are the ones telling us we need to go to another planet and use in situ resources, and you can't even eat Russian food, for crying out loud. Your suitcase is full of packaged, processed food."

The people who went over to Russia in the early days and made progress were the ones who weren't afraid to step outside their comfort zones—people, according to Barratt, "who didn't fear going to Russia and sitting down for hours of negotiation, inhaling cigarette smoke, being in poorly lighted rooms."

These people also had another skill that was critical to fostering respect and understanding: the ability to see the value that partners brought to the table, which requires taking a long-term view—including taking into account a partner's past. Such people, said Barratt, "were willing to treat these guys with the respect they deserved because of what they've accomplished." A partner's different perspective is valuable, but the very fact that it *is* different means that it will require work, humility, time, and resources to incorporate that perspective. At times, this will require checking one's pride at the door.

In the early days of the partnership, egos on both sides tended to present a huge barrier to collaboration. Both sides tended to see its way as the only way of doing business in space, and neither side fully understood why the other side was operating the way it was. The Russian design philosophy, for instance, is very different from that of the Americans. NASA tends to emphasize technology and redundancy rather than ruggedness, whereas, according to Bill Gerstenmaier, who is now head of all human spaceflight for NASA, the Russians design very robust systems.

> They couldn't understand a spacecraft that's so fragile that when you point it at the sun for a couple of hours, components overheat and quit working. . . . Many of the differences were because the U.S. was geared toward short-duration spaceflight, while the Russian operations were geared toward long-duration missions. At first we didn't realize why those differences existed. By working with each other, each nation learned the value of different approaches and of having an open mind.

At times, too, respect for the other's abilities requires a willingness to give up some control, to leverage the trust that has been built and empower partners to make real contributions toward shared goals. This, at times, can lead to someone other than us being placed in the "critical path," the position where the project would fail if the partner doesn't deliver.

Even today, recognizing this reality and giving up some control remains difficult for some in the United States, who tend to think of the ISS as an American space station with international cooperation, rather than as an international project with U.S. support and participation. The ISS would not be the incredibly capable orbiting research facility it is today without either Russians or Americans, just as it couldn't have been built without the Canadian robotic arm used in its construction. Nonetheless, according to Gerstenmaier, "It's hard for lawmakers outside NASA to see that dependency; they want to be in control. To be fully in control takes tremendous resources that are not available anymore. The only way we can succeed is to work together. The challenges of space exploration require us to work internationally and to work globally as a way to move forward."

Ego and arrogance of the kind seen in some of the early collaboration is perhaps the biggest impediment to seeing the value in others and therefore letting a partner take the lead where appropriate. If collaboration is to work, we need to realize that solutions can come from anywhere, and capturing the unique solutions arising from different perspectives means not dismissing things outright. It is vital to take the time to critically evaluate proposals and suggestions from valued partners.

ESTABLISH REAL, SINCERE RELATIONSHIPS

After mutual respect and understanding are achieved, it is possible to establish real, sincere relationships, which is the foundation of a solid long-term collaboration. But building these relationships is hard and takes time, so there should be some shared responsibility for the relationship, and some permanence to it. This will be facilitated if both sides strive to limit turnover.

Tom Marshburn summed up the importance of permanence: "Russia didn't move people out like we do. We try to capture our knowledge, put it on a piece of paper so the next person can do that same job. We had the mind-set that all knowledge can be caught on paper. The Russians didn't believe that—they put value on the individual who has the knowledge in his or her head and experience—something which just can't be written down and handed off to the next person. That affected our relationships quite a bit."

NASA's initial, transient approach made it much harder to establish the relationships necessary for true collaboration, and it soon became apparent that rapidly cycling people through the programs was not working. As Barratt put it, "Your signed agreement is only as good as your relationship with the person or the group." Therefore, NASA eventually shifted to a strategy that tried to maintain a more constant face in the programs in Russia.

One such face was John McBrine, who was present during the early days of the Shuttle–*Mir* program. He noted, "Things worked because I lived there. People saw me on a daily basis and I formed relationships with them. I didn't just come over for a week, demand stuff, and then leave. I was available to my Russian colleagues twenty-four–seven for a year. I think that goes a long way, because it enables us to see each other's perspectives."

McBrine said he still works with some of the same people at Star City that he worked with in 1994, but that hardly anyone he worked with at NASA in 1994 is still there. And, he observes, "This frustrates the Russians, because they are constantly asking, 'Who am I going to be working with next?'"

Gerstenmaier also saw his willingness to move to Russia and embed into Russian operations as critical to the effectiveness of the partnership. The Russians typically saw personnel come over for a couple of weeks and then leave, and the impression on the American side was that no one worthwhile would spend any significant time in Russia because there was nothing to be learned. "But when the Russians saw me stay there and leave my family back in the U.S., and understood that my motivation and love of human spaceflight was exactly the same as theirs—through that personal relationship, things grew. Then they kind of adopted me."

Today, the United States and Russia have long-term offices in each other's countries. Key space program personnel from both countries—along with their families—live full time in the other country for a couple of years or more, enabling a variety of informal interactions and thus promoting trust. In addition, a U.S. astronaut (or, at times, an astronaut from one of the non-Russian partner nations) resides and works in Star City as the full-time Director of Operations: Russia. The director facilitates the training and logistics of all U.S. and partner nation astronauts in Star City. Likewise the Russians have a similar position in Houston, called the Director of Operations: Houston. Today, the entire ISS program operates in a very integrated manner. Besides selecting astronauts from the various partner nations to command the ISS, astronauts from outside the United States have, at times, been placed in key leadership positions within the NASA Astronaut Corps.

Beyond these formal arrangements, impromptu social interactions have also served a critical role in fostering cooperation. Charlie Bolden, the commander of STS-60, the first space shuttle mission to include a Russian cosmonaut on the crew, was one of the first people in the Shuttle–*Mir* program to demonstrate the power of informal interactions to forge personal relationships. Bolden invited his STS-60 crewmates, including cosmonaut Sergei Krikalev, his backup, Vladimir Titov, and others, to dinner at his house. One of the other guests was Mike Foale, who would later fly with Titov on STS-63. Bolden gathered everyone in his living room and gave a heartfelt toast about flying together and getting to know each other. The evening left a lasting impression on both Foale and Krikalev.

These types of informal interactions, which continue to this day among all the ISS program partners, really laid the groundwork for the tough work ahead for the partnership. "What makes the partnership work is trust," Krikalev said. "I flew with Charlie in space. I know him well enough to know that when he says something, he means it." People are more likely to remember the great social interaction they had with a colleague than the great meeting they both attended.

DECISION MAKING IN THE TRENCHES

Ultimately, this kind of personal relationship benefited not just the U.S.–Russia relationship but all of the ISS program partners. Some of those people who worked hard in the trenches to form deep, lasting relationships are now in senior positions in the ISS partners' individual space programs, running the collaborative programs that they themselves started in collaboration with their counterparts, when all were at a lower level. In fact, during the initial space station partnership, a great deal of progress came from empowering workers at the operational level to make decisions and do their jobs. Barratt noted, "When we were free of our management, free to work directly with our counterparts in Russia and make decisions to move us forward, that's when things flourished. When we had to review every step with management—who went up the chain, who didn't understand, who hadn't made the kind of progress we had with our Russian counterparts—things went south."

As a collaboration becomes more mature, it can, at times, become more difficult to keep decisions at such a relationship level, because bureaucratic infrastructure starts to catch up with the process. NASA psychologist Al Holland noted, "I think the interactions with the Russians back then were a lot more frequent and productive. Things moved faster than they are currently, because you didn't have a big bureaucracy. . . . I'm sure from the top it looks good, because it keeps things organized and it fits in with contracts well. From my point of view, I much preferred just the direct interactions. I'd say now there's a fair gulf in interpersonal relationships, and we now rely more on agreements we've already worked out."

As a collaborative program grows and evolves, it's increasingly important to find the proper balance between institutional requirements and decision making through personal relationships and interactions. This balance must also take into account the different cultural norms the partners bring to the table.

SMALL STEPS BEFORE A GIANT LEAP

It is extremely unlikely that we would have the fully functioning international partnership that we have now if the Shuttle–*Mir* program had not come first, providing the two sides with experience working together without requiring commitment to any huge investments. And before Shuttle–*Mir* was the Apollo–Soyuz program, which did not require anywhere near the level of cooperation needed on the ISS or Shuttle–*Mir* programs. In describing Apollo–Soyuz, Bill Gerstenmaier said, "We essentially had to build a docking mechanism between the two capsules. We didn't have to share a lot of data, and we did that at the height of the Cold War, which was pretty symbolic."

Such incremental collaboration, however symbolic or simple, is nonetheless extremely important in laying the groundwork for larger projects. One of the biggest impediments to successful collaboration is trying to do too much, too soon. As the excitement of a potential collaboration builds, it's important that none of the partners get ahead of themselves. There is great value in taking incremental steps, in building the partnership deliberately and thoughtfully, enabling the partnership to survive inevitable trials and tribulations.

Beginning the U.S.–Russian partnership with a level of cooperation that seemed in some respects superfluous, for instance, initiated the initial discussions to explore deeper collaboration. Then, as the partners continued to work together, they were able to build an effective team to do progressively grander projects. Ideally, the ISS program will just be one more incremental step on an expanding, incredible journey of exploration and understanding, taking us higher and farther.

Mike Barratt believes that the collaborative effort on the ISS has made a trip to Mars as an international venture very possible. "Because of these years working together, I know exactly who to call, exactly who knows what. We know each other and we trust each other. There just has to be a background assumption by every-one that we can do it together, that it's our future. Otherwise, why invest? If we did all this work and stop at the ISS, what a waste."

A trip to Mars would be something big, valuable, and impor-tant that the ISS partnership and other nations could accomplish together, and one of the best ways to create a partnership that can weather inevitable storms is to create something so meaningful that the inevitable bumps in the road don't derail the partnership. All partners must create the reality that there's too much invested and too much on the table to risk losing it over the small stuff. It's not required that everyone have the exact same end objectives, but partners must share an understanding about the value of the relationship, regardless of any slight misalignment of goals. In terms of the ISS, for instance, Gerstenmaier noted, "Each partner probably has a slightly different motivation for what they're trying to get out of the station, and that's OK. But each individual's moti-vation is so strong that it keeps them all in the game no matter what comes."

LIVE UP TO YOUR COMMITMENTS

Building the personal relationships that fostered trust and friend-ship among the ISS partners took time and relentless effort. And a large part of that foundation of trust was made possible by the part-ners being there for each other, in good times and in tough times, and following through with what was promised. When the Soviet Union became Russia, for instance, the country's space program faced dire funding shortfalls. People were going without pay, and even essential supplies were unavailable. Al Holland's impression was that the Russians thought their partners might use the Rus-sians' hardship as an opportunity to take advantage. But the Amer-icans did not take advantage. Instead, said Holland, "We brought over to our counterparts whatever they needed, [which] caused the Russians to reconsider their relationship with us in a positive way."

Likewise, after the *Columbia* tragedy and the subsequent grounding of the U.S. space shuttle fleet, many worried that the Russians would take advantage of that situation, perhaps doing things like jacking up the price the United States would pay for Soyuz flights to the station. The reality, however, was that after *Columbia* there was not a sentiment of "us and them." Instead, there was just "us," one international space family all working together to get the program back on track and to get back to the business of space exploration. And the United States, for its part, made a commitment to recovery from the tragedy, not just for itself but also for its partners.

A couple of days after *Columbia*, Mike Foale and Alexander Botvinko, the deputy head of the spaceflight program for Roscosmos, went for a walk around the Baikonur tree garden.[1] Foale asked Botvinko how long he thought it would take the United States to recover, to which Botvinko replied, "I've seen this a number of times in various accident scenarios within our own programs. But don't worry, we have the Soyuz. We're going to maintain the partnership." Foale felt that the Russians were everything the United States could want in a partner, and at that moment, the two countries had arrived at a true lasting partnership.

During the period immediately after the *Columbia* tragedy, I too felt the solidarity of the Russians and our other international partners. We mourned and grieved together, and through that dark period we became something more resembling family. We came to realize that being able to count on each other's support when it's needed is more important than trying to agree on things all the time. That is true collaboration.

PART II

LOOKING EARTHWARD

CHAPTER FOUR

One Moment in Space

✦ I lay on my back, surprised at how calm and focused I felt, strapped to four and a half million pounds of explosives. Years of training, preparation, hard work, and prayer had led me to this moment, on May 31, 2008. Although I was a trained and qualified shuttle pilot, I was flying on this, my first flight, as Mission Specialist 2 (MS2), the flight engineer. My journey to this seat had begun about a year and a half earlier, when the chief of the Astronaut Office, Kent Rominger, sent a note to all the pilots in my astronaut class asking if we would consider undertaking our first flight as a mission specialist. I told him I would be happy and honored to fly wherever needed, which apparently was the correct answer. Around a month later I was penciled in for my first spaceflight assignment. As MS2, I was seated between and behind Commander Mark Kelly and Pilot Ken "Hock" Ham, and to the left of Karen Nyberg, who was MS1.

Sitting there, I felt some apprehension, of course. But I was also reassured by the idea that what we were about to do would make a contribution to humanity and, at this point, that the outcome of the launch was largely out of our hands. The primary mission objective of STS-124 was to deliver the central pressurized laboratory of the Japanese Kibo ("Hope") module and its associated external robotic arm to the International Space Station. The Kibo laboratory is the largest pressurized module on the ISS and includes many stations for scientific research. It now also includes an exposed research facility—the "porch"—which is fully exposed to space.

Just as the launch countdown was about to resume, after a planned hold at T minus nine minutes, launch director Mike Leinbach radioed, "Okay, Mark, it's a gorgeous day to launch. So on behalf of the KSC [Kennedy Space Center] launch and processing teams, I'd like to wish you good luck and Godspeed as you deliver the next piece of Kibo to the International Space Station."

Commander Kelly pulled the radio transmit button on the joystick of the Space Shuttle *Discovery* and said, "Okay, thanks. You're right. We're going to deliver Kibo, or 'hope,' to the space station. And while we all tend to live for today, the discoveries from Kibo will certainly offer hope for tomorrow. *Mina-san arigato, itte kimasu* [Thank you, everybody, we're going now]. Now, stand by for the greatest show on Earth."

EIGHT AND A HALF MINUTES TO SPACE

Hours earlier, Mark, Hock, Karen, and I, along with the three other crewmembers of STS-124, Greg "Taz" Chamitoff, Mike Fossum, and Aki Hoshide from Japan, had departed from crew quarters at Kennedy Space Center, waved to those Space Center personnel who had gathered outside the quarters, and boarded the "Astrovan," part of a large convoy that also included a SWAT team and a helicopter gunship. When we asked Mark about the purpose of the gunship, he joked, "That's in case we change our minds and turn around." The convoy got smaller and smaller as we got closer to the launchpad and our van made several stops along the way to let off management personnel. By the time we arrived at the launchpad, the convoy included only the van, carrying just our crew and a couple of technicians.

That afternoon, the launchpad was a sight to behold. It usually is bustling with activity, with hundreds of technicians, engineers, and support personnel poring over every square inch of the rocket and spacecraft. On the day of the launch, however, the entire launch complex was deserted except for a few folks on hand to help us get strapped in and to ensure that the spacecraft was configured properly for launch. The spacecraft looked alive, massive and shimmering in the sun, white clouds of oxygen rolling off the rocket in an eerie silence.

Before boarding the launchpad elevator, we all took a few moments to stand in the beautiful Florida sunshine, enjoying a cool breeze coming off the ocean, in awe of the tremendous human accomplishment that the space shuttle represented. Then we all got on the elevator and rode it to the 195-foot level, where the gantry that provides a bridge to the space shuttle was located. Also on the 195-foot level, a sign announced, LAST BATHROOM ON EARTH.

After making use of the last bathroom on Earth, I crossed the metal-grated gantry over the long drop to the concrete launchpad below. I had made the trip across the gantry many times before, but this time it was different—this time it was for real. At the end of the gantry was a small workroom called the White Room, which provided access to the spacecraft and a working area for technicians to help us get suited up. One after the other, we entered the White Room, got suited up, crawled through the hatch, and slid into our seats, allowing ourselves to be strapped in by ground personnel. When it was my turn in the White Room, technicians helped me don all the equipment I would need and did some final checks. I then crawled into *Discovery* and up onto the flight deck—although, in the launch position, it was more like right and *down into* the flight deck.

Now, as we lay on our backs, approaching the moment of liftoff, I thought about the thousands of people lining the roads to watch our launch. As I realized that we were really going to go, and that a cancellation of the launch was unlikely at this point, I felt relieved that my friends and family wouldn't be disappointed. I also wondered briefly what I was getting myself into.

The digital clock in the center of *Discovery*'s control panel, right in front of me, counted down to within a minute. As it counted below ten seconds, I prepared for the main engines to light. At six seconds remaining, a low rumble started, which very quickly built in intensity. I watched as the digital engine display on all three main engines rose to one hundred percent. Hock said very calmly, "Three at a hundred. Hang on."

Using my wrist mirrors, I looked briefly through *Discovery*'s overhead windows and watched towering white smoke billowing behind us against the backdrop of the Florida coast. As the engines came up to speed, we rocked forward and then back. The solid rocket boosters fired just as the spacecraft rocked back to its original position and the clock counted to zero. My wristwatch read 5:02 p.m. EST.

During launches seen on television, or even in person, the shuttle seems to lumber slowly off the pad. Sitting inside the vehicle, however, it feels anything but sluggish. It feels more like the entire shuttle has just been released from a giant slingshot, leaving no doubt in anyone's mind that we are going somewhere, and fast. The launchpad passed quickly below us as we leaped into the sky. Light streamed into the cockpit as *Discovery* rolled to put us on the correct trajectory to rendezvous with the ISS.

Thirty-six seconds into the launch, the engines were throttled back to slow our acceleration and ensure that we would not exceed the limits for dynamic pressure. This happened as we accelerated through 80 percent of the speed of sound, traveling 609 mph. I made the "ninety seconds" call, indicating that, if needed, we could now take manual control of the spacecraft.

As we passed an altitude of fifty miles, Mike called over the intercom from the middeck, "You have three people on the flight deck that need to do a big high five." Fifty miles is the altitude at which first-time fliers become certified astronauts. The first-time fliers on the middeck, Aki and Taz, had already exchanged their high fives. On the flight deck, Karen, Hock, and I, now officially "flown" astronauts, also happily complied.

At eight and a half minutes into the launch, the engines abruptly shut down, according to plan. Inside the spacecraft, we all instantaneously transitioned from being pushed into our seats at a force of three times our weight to being completely weightless. At this point I thought, "We made it," by which I meant "We survived." I also was acutely aware that my childhood dream of flying in space had just come true.

A WINDOW ON THE EARTH

A little over an hour after launch, when we were all safely in orbit, I was finally able to unstrap and attempt flying around in zero gravity for the first time. Weightlessness was wonderful, and I was surprised at how natural it felt. I was relieved that I wasn't nauseous at first, perhaps because my body had a previous context with which to compare this new environment. I had experienced zero gravity many times, flying high-performance fighter aircraft. After being in this new environment for a few hours, though, internal alarms went off, my body telling me, "Hey, zero-g is not supposed to last this long. Something must be wrong." During my first few hours in space, I did periodically feel nauseous, but it was really just a minor inconvenience and didn't deter from the work that had to be done to turn our rocket ship into an orbiting spacecraft.

My first task after unstrapping was to open *Discovery*'s payload bay doors. Karen and I floated over to a control panel in the aft part of the flight deck. Above the control panel, overhead windows gave us a sneak peek at the grandiose sight we were both about to experience for the first time. After we went through a series of checklist steps, switch throws, and inputs to the computers, *Discovery*'s massive payload bay doors opened, revealing a window onto Earth.

To say the view was breathtaking would be an understatement. The first thing that struck me was how thin the atmosphere appeared, as if the entire planet were wrapped in a paper-thin blanket. And in that moment, I was hit with the realization that this delicate layer of atmosphere is all that protects every living thing on Earth from perishing in the harshness of space. So, although this was an incredibly overwhelming visual experience, it also was much, much more than just a visual experience. For me, it was a profound feeling of detachment, with a simultaneous connectedness. I felt a visceral, physical separation from the only world I had known since birth, while at the same time I was able to see that world with my own eyes.

Years later, back on Earth, the words of another person with a similar experience helped me to understand the deeper implications of the experience. My friend Wasfia Nazreen is on an amazing quest to summit the highest peaks on all seven continents. She embarked on the quest on the fortieth anniversary of the birth of Bangladesh as a country, beginning by summiting Mount Kilimanjaro, in October 2011. The Bangladesh on Seven Summits campaign seeks to raise awareness for women's rights in Wasfia's home country, celebrates women's strength and progress over the past four decades, and calls upon the youth of Bangladesh to take on responsibilities and new challenges to move the country forward.

In discussing climbing Mount Everest, Wasfia wrote, "On the summit of Everest, the overriding feeling was that of gratitude, which literally made me feel I was connected to every single human being out there, though my heart in reality crawled all alone. . . . Sitting on the top, I felt tinier than a bug leeching on the expansive stretch of these Himalayan gods and goddesses. And an intense amount of respect and love flowed through me for the courage and dedication my ancestors have gifted me."[1]

What really struck me in Wasfia's description of her experience was that overriding feeling of gratitude, which in some way connected her to every single human being. This was exactly what I experienced in space: immense gratitude for the opportunity to see Earth from this vantage, and for the gift of the planet we've been given. In some way I can't explain, being physically detached from Earth made me feel deeply interconnected with everyone on it. I distinctly remember this feeling of profound thankfulness, but Wasfia's words helped me to process that awareness of inter-connectedness with the inhabitants of the planet.

You can't help but fall in love with the beauty of Earth. Up on the shuttle, Karen and I just floated for a moment and took in this colossal view. But we still had much to do, so we allowed ourselves only a simple glance and a smile, acknowledging that we had just become two of the few hundred people in the history of humanity to experience this view.

"YOU GUYS LOOKING FOR A PLUMBER?"

After a few hours of busy activity, everyone was finally ready to end a very long, exciting, and memorable day. That night was my first in a world without gravity, where all the rules that I had followed my whole life—silly rules, like you can't sleep on the ceiling—were thrown out the window. We staked out spots on the floor, walls, and ceiling, attached our sleeping bags, and called it a night.

The sheer exhaustion from all the exciting events of the day (and the sedative side effects from a shot of the antinausea medica-tion Phenergan) allowed me to drift off to sleep very quickly, and it turned out to be the best night's sleep I had for the duration of the two-week mission. But I still felt a little strange before I actu-ally fell asleep. I wasn't quite sure what to do with my head. I had no pillow, so my head just floated there awkwardly. It wasn't until three years later, after many weeks in space and after the muscles in my neck adjusted to this strange environment, that I was finally able to sleep comfortably without gravity.

During space shuttle missions, we don't keep track of the days of the week; we keep track of flight days. And because NASA abbreviates everything, each day was referred to as an FD, with FD 1 being our first full day in space. So FD 1 and FD 2 were spent pre-paring our ship for rendezvous and docking with the International

Space Station. We performed periodic maneuvers to gradually align our orbit with and move in closer to the space station. I also spent a good deal of time with my spacewalk partner and mentor, Mike Fossum, ensuring that our space suits would be ready to go as soon as possible after we docked with the ISS. Everyone awoke excited on FD 3, the day we would dock to the space station, which was to be the busiest day of the mission.

As we approached, the ISS looked like a bright star, but as we moved closer, the giant structure took shape. The ISS is massive, bigger than a football field, and on Earth it would weigh more than one million pounds. The facility is made up of different modules, built by the different nations of the International Space Station partnership. Each module has a specific purpose, providing crew quarters, laboratory facilities, life support, exercise equipment, a galley, and of course, strategic places for observing Earth. And now we were going to provide the largest addition: Kibo.

We arrived six hundred feet beneath the space station and executed a rotational pitch maneuver, or RPM, which involved doing a 360-degree flip under the space station so that the ISS crew could take very detailed technical pictures of the thermal protection tiles on *Discovery*'s belly. After completing the RPM, Mark flew *Discovery* up in front of the space station and slowly backed in for a picture-perfect docking. Amazingly, after traveling almost a million miles to catch up with the complex, our tolerance for docking was only plus or minus three inches. NASA astronaut and ISS crewmember Garrett Reisman rang the ship's bell on board the station and announced, "*Discovery* arriving."

There was still a great deal of work to be done, but eventually everything was completed and the air pressure was equalized between *Discovery* and the space station. As we opened the hatch, Mark called across to the space station, "Hey, you guys looking for a plumber?" We were carrying spare parts for the space station's broken toilet. On the other side of the hatch was the smiling face of Garrett, happy to see the parts to fix the toilet, but also happy because, after he had spent three months in space, his ride home had arrived.

We all floated across to the space station and were greeted by Garrett, Russian cosmonaut Oleg Kononenko, and ISS commander Sergey Volkov. We had a few moments to say hello to our friends before we started a safety tour of the space station, to get accustomed to this massive facility and to review the procedures to follow in case of a fire, air leak, or release of a toxic substance.

As soon as the tour was over, Mike and I hurriedly transferred all of our gear and suits into the station's airlock, readying everything for our spacewalk the next day. We spent our first night on the station in the airlock, at reduced pressure, to purge our bodies of excess nitrogen and lessen the chance of getting the bends when we went out the door the next day. That night, Mike and I established a routine that we would follow over the course of three years and four spacewalks together: After all our tools, gear, and spacesuits were all set up for the spacewalk, as we were winding down and setting up our "campsite" for the night, we listened to Pink Floyd's *Dark Side of the Moon*. The next day, as we were making final preparations prior to going out the door, we would switch the mood and listen to Led Zeppelin. The specific Zeppelin songs didn't matter, but we were required to listen to "Kashmir" before heading out into the vacuum of space.

"DON'T LOOK DOWN"

On my fourth day in space, after listening to the obligatory "Kashmir," we suited up to head out the door. I was filled with both excitement and apprehension. The hectic pace of the mission, along with my inability to figure out how to sleep comfortably, had left me in a state of significant fatigue. I was so tired I was not sure I would be able to accomplish our tasks.

After Mike and I were both buttoned up in our respective space suits, Mike discovered a problem with his communications system, which meant a slight delay while Mission Control figured out what was going on. This situation gave me about fifteen minutes of downtime and provided an opportunity for a micro-nap. Amazingly, I was so tired I was able to fall asleep right before undertaking the most incredible thing I could ever imagine!

Once the communications problem was fixed, Mark, Garrett, and Taz stuffed Mike and me into a part of the airlock called the crew lock. Mike went in first, head first, followed by me, feet first, followed by all the equipment that we would take out the door with us. Eventually we were packed in the crew lock like sardines, with barely room to move. Mark closed the hatch, sealing us off from the rest of the space station, and we waited for about thirty minutes while all the air in the airlock was vented either back into the ISS or out into space. When the air pressure reached zero, Mike opened the hatch, and reflected sunlight from Earth's surface flooded the airlock.

After all the checks were complete, Hock, who was acting as choreographer (or, in NASA-speak, the intravehicular officer, or IV) of the spacewalk, radioed, "OK, boys, you ready to rock and roll?" Mike responded, "We're ready to rock and roll," and with that, he headed out the door. I handed our equipment out and then followed. Mike, having done three previous spacewalks, felt the need to give me some words of advice: "Ronnie, don't look down" was all he could muster.

Well, I did look down, and up, and every direction I could. And I was in awe.

I really had a hard time processing this new existence, which seemed filled with mixed messages. Part of my brain was in awe of the beauty of the scene, was aware that this was the most amazing, special, and exhilarating thing I'd ever done, but another part of my brain was unable to fully process that what I was doing was actually real. It was almost as if I were watching everything unfold before me as in a movie, and I was detached from it. Soon enough, though, reality sank in, and the sheer number of tasks that lay before us made me quickly abandon my debate about the nature of reality and get to work.

In the vacuum of space, we experienced extreme temperature shifts and either a sunrise or a sunset every forty-five minutes. When the sun sets, the entire space station is cast in darkness, except for the sporadic outside lights of the station and the lights on our helmets. The temperature in the shade falls to –250°F, while in the sunlight the temperature climbs to more than 250°F. Spacesuits, of course, are basically self-contained spaceships with temperature control, so we are not overly cold or warm during spacewalks.

This was the environment of our construction work site, and at times I really did feel like I was on a 240-mile-high skyscraper construction site. From the top of the main ISS truss, which supports the station's massive solar arrays, the scene reminded me of a photograph I once saw of construction workers sitting on a beam, eating lunch high in the air during the construction of the Empire State Building. In our case, however, we were a lot higher, and the whole job site was moving at five miles per second. My initial apprehension about my level of fatigue turned out to be for naught, because the sheer exhilaration of the experience provided ample motivation to get all our assigned tasks completed—and then some.

THE ORBITAL PERSPECTIVE

Over the course of our 8 days, 17 hours, and 39 minutes docked to the space station, our space shuttle crew and the ISS crew transferred tons of equipment and cargo, installed and activated the Kibo laboratory, and did some required maintenance on the orbiting outpost. Mike and I spent a total of 20 hours and 32 minutes outside. On our third and final spacewalk of the STS-124 mission, we were tasked with replacing a five-hundred-pound nitrogen tank. I was to remove the old tank, carry it to another part of the station, swap it for a new tank, which Mike would have ready, and then carry it back to where it needed to be installed on the station. Carrying such a heavy load in space is technically possible, because the object, like everything else, is weightless. However, the tank was very bulky, making it very difficult to drag from one part of the space station to another. It also would be likely to bang into equipment on the station along the way—not a good idea. To get around this difficulty, we used the space station's robotic arm.

The Canadarm2 is an amazing engineering accomplishment that served as a sort of construction crane to assemble and connect the various ISS modules. On our mission, the arm was used to reach into *Discovery*'s payload bay, lift out the massive Kibo laboratory, and attach it to the ISS. The main part of the arm is 57.7 feet long, and when a foot restraint is attached to the end of it, astronauts can be carried (or "flown") from one part of the station to another while carrying equipment.

I wedged my feet into the restraint and grabbed the handles on the nitrogen tank. With Karen and Aki operating the arm from inside the space station, the nitrogen tank and I were backed away from the main truss segment. The arm was then flown through a maneuver we called the Windshield Wiper, which took me through a long arc across the top of the space station. At the top of this arc I was about one hundred feet above the space station. The orbiting complex was on the dark side of Earth, and it was a little eerie to be rising above the lights of the space station and into the darkness above.

Approaching the top of the arc, I let go of the nitrogen tank with one hand and switched off my helmet lights, allowing my eyes to adjust to the darkness. When my eyes did adjust, I watched the universe unfold before me. The pitch-blackness of space slowly transformed into an amazing heavenly light show. I was no longer looking out into stark nothingness; I was looking out into infinity. I could see the Milky Way and even satellites occasionally whizzing by—the latter were visible because their higher orbit kept them in sunlight, which was being reflected in my direction.

After I traded the old tank for the new one from Mike, the arm flew me back across the same arc, back to my original position on the space station. This swoosh of the Windshield Wiper, however, occurred on the daylight side of the orbit. Now, at the top of the arc, I found myself looking down at the enormous International Space Station, one hundred feet below me, against the backdrop of our indescribably beautiful planet 240 miles below.

The sheer beauty of the scene took my breath away—but even more compelling than the beauty was the realization of the tremendous human achievement that the International Space Station represents. It is not only an amazing technical accomplishment—probably the most complex structure ever constructed—but also one of the most amazing examples of international cooperation. As I hung there with this massive box in my hands, looking down at the station against Earth, I marveled that fifteen nations, some that have not always been the best of friends, had found a way to set aside their differences and achieve something amazing in space. I wondered what the world would be like, and how many fewer problems we would all face, if we could figure out how to have the same level of cooperation and collaboration in our interactions on Earth's surface.

After we were safely back in the station, I had a few moments alone on the middeck of the shuttle, and the enormity of the previous few days really hit me. For me, with three spacewalks behind us, the most challenging part of the mission was over, and my relief that everything had gone well and all tasks were accomplished was palpable. As I floated there and took this all in, a wave of intense gratitude washed over me—the same kind of gratitude that my friend Wasfia talked about feeling on top of Mount Everest. I was grateful for the opportunity to have this experience and thankful for all those who helped and supported me along the way, especially those who kept me and my crewmates in their prayers. It seemed as if I could actually feel the prayers of my family and friends, including communities of cloistered Carmelite nuns in

Texas and Rome with whom I had become acquainted over my years of travel. This feeling of intense gratitude was one of the most moving experiences of my life.

After our time on the space station was complete, we counted noses to ensure we had all the correct people on the correct sides of the hatches and then undocked from the station. We left Taz on the space station and had Garrett with us for his return to Earth. After undocking, Hock backed *Discovery* away from the space station and flew a complete loop around the ISS so that we could take pictures documenting the condition of the station's exterior.

As we backed away, Garrett asked, "Are you guys ready for some theme music?" He had set up his iPod with external speakers on *Discovery*'s flight deck, and as we backed away he played "The Blue Danube" by Johann Strauss, perhaps best known from the soundtrack of *2001: A Space Odyssey*. During one part of our maneuver we were directly above the station, looking down at Earth framing the ISS at night, with Shanghai aglow in the background. I wished I could freeze this moment in time. Thanks to Garrett, when I listen to "The Blue Danube," all the memories flood back.

EVERYONE CAN HAVE AN ORBITAL PERSPECTIVE

I would like to be able to tell you that when I went to space I had an epiphany, that I looked at the Earth from space and was filled with an instant realization and given an orbital perspective—a perspective that we are all interconnected and that we're all riding through the universe together. But it didn't really happen that way. Before I ever launched into space, I already had a theoretical understanding of the orbital perspective. Going to space only confirmed what I already knew. It provided a concrete reinforcement of what I somehow intuited about the orbital perspective.

And while I do believe that the world would be a better place if everybody went to space and experienced the orbital perspective firsthand, you don't have to be in orbit to espouse an orbital perspective. It doesn't require going to space to realize that we are all living on a precious, fragile planet, that we are all in this together, and that together we have a great number of challenges to overcome. It shouldn't require us being in orbit looking at our planet from space to realize the sobering contradiction between the beauty of the planet and the unfortunate realities of life on our planet for the majority of its inhabitants. Adopting an orbital perspective simply involves incorporation of a big picture and long-term awareness of the full effects of the decisions we make, and an ability to put yourself in another's shoes. It requires that we develop empathy for the struggles that our crewmates on Spaceship Earth endure.

I believe the vast majority of the people on our planet want to make a difference and want to make the world a better place, and that most of the inhabitants of our planet feel a responsibility to leave the world better than they found it. So, if this is true, why do we still face so many problems?

It turns out that many of the causes of conflict and lack of cooperation here on Earth are in fact the same problems the United States and Russia experienced in the beginning of what became the ISS partnership, which we saw in part 1. Those two countries, along with the other ISS partners, were able to overcome these obstacles and create a massive technological achievement that serves as an example to help us to understand the orbital perspective. In the next two chapters we will look at exactly what the orbital perspective means and how that perspective can aid in solving the biggest problems here on Earth.

CHAPTER FIVE

The Orbital Perspective

✦ There have been many attempts to describe the shift in perception that astronauts often experience when viewing Earth from orbit or the moon. Perhaps the most comprehensive is Frank White's book *The Overview Effect,* which was first published in 1987.[1] In the first edition, Frank interviewed sixteen astronauts about their experiences in space, and he coined the term "overview effect" and developed it into a theory to describe the perspective shift that some of these space travelers experienced.

If the overview effect is a change in perception that one gets from physically seeing Earth from space and in space, then the orbital perspective *derives from* that experience and drives what we do with it. The orbital perspective is the call to action that results from the overview effect. The key for me is that the orbital perspective involves taking action. Seeing Earth from the vantage point of space not only provides a unique perspective but also can trigger a response that leads to a desire to make the world a better place for all its inhabitants.

FRAME OF REFERENCE

White isn't the only one to ponder the effect of space travel on astronauts' perceptions. One of the most frequent questions I'm asked about my experience living and working in space is what it is like to see Earth from space. The only way I can describe this rarest of human experiences is to compare the beauty of the experience with the beauty of experiences we find on the surface.

When we sit on a beach and look at a beautiful ocean scene, we may be a long way from our focal point on the ocean, but we remain physically connected to that focal point through the beach and the intervening water. When we stand at the rim of the Grand Canyon we are physically connected to the vast beauty of that scene, and we are anchored to the scene by gravity, which pushes us down into it. But when I looked at Earth from space I was physically detached from the beauty I was experiencing. Separating me from that immensity was a void, a vacuum, a gulf of nothingness. I also was floating effortlessly above (or below, depending on my perception) the beauty I was witnessing. I was not perceptibly connected to the scene by gravity or in any physical way.

Scenes on the surface of Earth can be very still, which is not true when one is orbiting the planet. A snowy winter scene might impress us with its quiet stillness, even though the ground we are standing on is rotating at 1,000 mph, while Earth rotates around the sun at 67,000 mph, while our solar system rotates with the rest of the Milky Way galaxy at 515,000 mph, and the galaxy is hurtling outward at 1.4 million mph. All that motion is outside of our awareness. But when I gazed at Earth from space, there was no escaping the fact that I was traveling across the surface at five miles per second—17,500 mph. The lighting, colors, and the motion of the scene were constantly changing. And when my brain was able to extrapolate the curvature of Earth, I was filled with the certainty that I was witnessing a planet hanging in the blackness of space. Earth became my ever-present travel companion. This is the overview effect, which makes possible the orbital perspective.

ELEVATED EMPATHY

In addition to the overview effect, however, there is another element to the orbital perspective, which I call elevated empathy. This concept really became concrete to me one night in July 2011, four months into my five-and-a-half-month mission on board the ISS. Prior to calling it a night after a typically long workday, I flew to my favorite spot to partake of my favorite space-based pastime: snapping photos of Earth from the cupola, an Earth-facing, windowed observatory on the bottom of the space station.

My goal that night was to take a sequence of still images of Earth for a time-lapse video project I was working on. To ensure that I had the correct camera settings, I took a few practice shots, and as I was reviewing some of these, one in particular caught my eye. I noticed a long, illuminated line snaking hundreds of miles across a large landmass. Initially, I wrote it off as the reflection of moonlight on a river, but it turned out to be not a natural reflection at all. That line was actually the illuminated human-made border between India and Pakistan.

Seeing that border from space, a true barrier to collaboration, had a huge effect on me. For the fifty-plus years that humans have been flying in space, astronauts and cosmonauts have commented on how beautiful, tranquil, peaceful, and fragile our planet looks from space. These are not trite clichés; it truly is moving to see our planet from space. But looking down and seeing a border between India and Pakistan, recognizing the undeniable and sobering contradiction between the staggering beauty of our planet and the unfortunate realities of life on our planet for many of its inhabitants, also inspires empathy for the struggles that all people face.[2]

This elevated empathy is an important aspect of the orbital perspective. Elevated empathy helps us realize that we are all riding through the universe together on this spaceship we call Earth, that we are all interconnected, that we are all one human family. This scar on the otherwise beautiful landscape was a compelling call to focus on the need for global collaboration to overcome the world's problems, to recognize that, in spite of our disagreements, we should behave like a family, should communicate, support, stand by, and care for each other.

And from this photograph, the orbital perspective was born. If I heard of a natural disaster or a famine or conflict on the ground, I could see those areas as we flew over them, and even though I couldn't see the details of what was happening on the ground, I could still feel connected to those circumstances. Mike Foale said, "What being in space makes you realize is that there are other people in the world." He stressed that this was the most important part of the experience. "You go over all those other places that you never thought about. You wonder, What's down there? What's life like down there? And as soon as you start that process you're already on your way to being a citizen of the world rather than just your country."

Even those who have never been in space can integrate an orbital perspective by keeping in mind the important ground-level details of a situation while zooming out as far as necessary to incorporate all factors that affect and are affected by the situation. This big picture view is achieved by pulling back not just in a spatial sense but also in a temporal sense, and this enables us to realize that the decisions we make and actions we take today will have a long-term effect on the trajectory of our global society. Many amazing, talented people have the correct pieces of the puzzle, but it's not until we pull back to the orbital perspective that we see how those pieces fit together.

CONFLICT ON THE SURFACE

During my half year in space, I was moved by many other views of Earth from the windows of the space station besides the India–Pakistan border. One such scene occurred while we passed over the Mediterranean Sea. I was struck by the contrast of colors, the dark blue of the sea against the bright tans and reds of northern Africa, all framed by the curvature of Earth and its paper-thin atmosphere. I quickly grabbed a camera and started snapping away. As it happened, Libya was framed within the picture, which I took on August

24, 2011, the day that Tripoli fell during the Libyan Civil War. Looking at the photo, I found myself thinking deeply about what was occurring at the microscopic level within the frame of this beautiful scene. The violence and suffering that was imperceptible from the orbital perspective was all-encompassing when you zoomed into the details of the two-dimensional plane of the conflict.

Most of the many conflicts raging around the world involve complex factors of history, ethnicity, class struggle, poverty, religion, and so forth. Usually, these conflicts adhere to the ideology of scarcity; actions tend to be based on a fear that the other side might attempt to take control of material things such as water, land, oil, or some other resource. But for the most part, these conflicts are, at heart, two-dimensional conflicts. There exists an "us" and a "them," and both sides tend to demonize and dehumanize the other. If one side gains, the other side loses.

When we pull back to the orbital perspective, however, we see the field of conflict in a new light. We see combatants engaging in violence against each other in a small slice of an apparent paradise. Each side in an armed conflict tends to diminish the humanity of the other, but the orbital perspective provides the view that both sides are fully human, and that by degrading the human dignity of one we degrade the human dignity of all.

The currency of the orbital perspective draws upon the unlimited resources of respect, compassion, empathy, and creativity. If I offer compassion or share an idea, it doesn't lessen my supply of creativity or empathy. Those involved in conflict usually see their opponents as completely separate from themselves, but if those combatants could pull back to the orbital perspective and employ some elevated empathy, they would discover the unity and interconnectedness they were missing. If they were able to look at the long-term effects of their actions, they would see there's a better way forward.

This is really no different from the joint interests that led the United States and Russia to collaborate in space, or that enabled astronauts and cosmonauts to work together for their common survival in the face of fire or cabin depressurization. Pulling back to the orbital perspective helps us to realize that, like those space travelers, we're all riding through the universe together. There are no passengers on Spaceship Earth, only crewmates, and we all have a responsibility to mind the ship and take care of our crewmates. In fact, when we pull back temporally and look at long-term effects, it becomes apparent that we *must* collaborate, for the survival of life on Earth.

LAND OF A THOUSAND HILLS

In space, I was hit with the sobering contradiction between the beauty of Earth and the unfortunate realities of life, such as conflict that exists on our beautiful planet. But understanding this contradiction does not require a spacewalk or even a trip to space. The overview effect is, in some sense, available to us all. We've had pictures of Earth from space for decades, and today we even have tools such as Google Earth, which enables anyone with access to make a virtual trip around the world and then zoom in to any detailed portion of it. And elevated empathy is likewise accessible to anyone who reads the newspaper or has firsthand knowledge of local problems and allows himself or herself to feel some connection with those problems.

I've seen this contradiction between global beauty and earthbound conflict from space, but I've also been confronted with it on the ground. In 2006, I was giving a presentation about space to a group of students in rural Rwanda, "Land of a Thousand Hills." Since suitable facilities for a meeting like this can be scarce in such areas, we decided to open up a chapel on the grounds of Mugonero Hospital, where I was helping to install solar panels. This chapel had been closed off since the Rwandan Genocide in 1994, which claimed the lives of almost one million people. As I was about to begin my presentation, I noticed bullet holes and what appeared to be bloodstains on the concrete walls of the chapel. On April 16, 1994, approximately two thousand men, women, and children who sought refuge in this chapel and the surrounding hospital grounds had been slaughtered here.

The location was somber and the events horrific, but as I interacted with the students during the presentation, I was struck by their enthusiasm, attention, optimism, and intelligence. These students and their families had all been affected in one way or another by the genocide and the extreme poverty of the area. But in this moment of awe and wonder, while being exposed to space exploration, they rose above their circumstances, rose above the horror of their nation's recent history, and realized that a bigger world exists. For a moment, they saw an additional dimension of their universe contained within the beautiful rolling hills and lakes of west Rwanda.

I encountered this beauty again as I later interacted with children of the nearby L'Esperance orphanage, not inside a structure that had seen a violent past, but outside, on a hilltop on the orphanage grounds, overlooking the beautiful Lake Kivu. While I was there, I attended a service at a nearby church, where I was absorbed by the enthusiasm, lively music, and colorful dress. There was a distinct feeling of optimism in the air.

During the service I noticed what appeared to be a painting of the Eucharist contained within a monstrance centered directly above the altar.[3] As I examined this "painting" a little closer, though, I realized it was not a two-dimensional painting but a three-dimensional imprint in the wall. Suddenly, everything came into clear focus. I was actually looking at an impact crater made by a rocket-propelled grenade impact. This church, too, had been the site of a massacre during the genocide. While rebuilding the church, the parishioners apparently saw the violent scar centered directly above the altar as a symbol of hope, a sign that they could rebuild from the ashes and that there was more out there than the two-dimensional violence of their past.

THE WORM'S EYE VIEW

In order to dive a little deeper into what the orbital perspective means, I'll borrow a term from my friend, and one of my personal heroes, Nobel Peace Laureate Muhammad Yunus. Professor Yunus describes a perspective he calls the worm's eye view. This concept stemmed from his desire to step beyond the lofty, big picture economic theories he was teaching at the university and to get down to the ground level and experience the details of what it's like to live in an impoverished village in Bangladesh. He wanted to go from the lofty bird's-eye view, where the landscape and the economic and societal features could be seen, and dive down to see the intricate details of all the factors affecting poverty.

Professor Yunus found that the worm's eye view differed sharply from the top-down theories he was teaching. The worm's eye view revealed impoverished people who were enterprising, resourceful, and creative but were being held down primarily by circumstance such as loan sharks, corruption, or not having access to basic services. Taking this worm's eye view, Professor Yunus created microfinance and social business movements that have helped to lift millions of people around the world out of extreme poverty.[4]

Professor Yunus stressed that the bigger an organization grows, the more distant it gets from the ground level and the real issues in its client communities. "All you have is numbers, and you forget about the people."[5] However, he does not suggest that we should abandon the bird's-eye view entirely. Instead, he argues that both perspectives are important. Before seeking what Professor Yunus calls a "global view," we should first be rooted on the ground and understand the details of a situation on the micro level.

The orbital perspective is a similar blending of the small and big pictures. The orbital perspective and the worm's eye view are two sides of the same coin that, when synthesized, can manifest an elevated empathy, which can lead to a feeling of connectedness and oneness with those suffering from the worm's eye details on the ground. Being detached from Earth, for me, made this synthesis seem obvious, but being detached from Earth is not a requirement to manifest elevated empathy. The orbital perspective allows us to see things on a global scale, simultaneously considering the worm's eye details while pulling back past the bird's-eye view, past the landscape and terrain, and all the way out to a perspective that allows us to see all the pieces of the puzzle, how they connect, and the picture they paint for our global society. We must adopt a perspective that incorporates the best that both the worm's eye view and the orbital perspective have to offer.

LIVING IN THE WORM'S EYE VIEW

Being able to see the details afforded by the worm's eye view is important, but living with an entirely worm's eye view poses some problems. Living in the worm's eye view means that we are focused on our own immediate sphere in a very narrow and short-term manner. We think about our personal needs or our family's needs. If we're altruistic, we may also consider the needs of our neighborhood, our city, or our region. We may think about the plight of others throughout the world, and we may even donate some money to the many efforts to address these needs. There's nothing wrong with any of these things, but each paints a very incomplete picture.

For instance, during the early days of the modern environmental movement, the term *NIMBY* ("not in my backyard") was coined to describe people who voiced concern, and at times outrage, only about things that were happening in their immediate vicinity. But the orbital perspective shows us that your backyard is much bigger than you think—the environmental radius of effect is actually global.

Similarly, the worm's eye view of "doing good" might consist of donating money to a charity—and this may produce some good in the short term. But pulling back to the orbital perspective may reveal a different long-term, big picture view. Perhaps helping in one area creates problems in another. What if short-term assistance actually makes a situation worse in the long term?

The worm's eye view would attempt to solve a specific problem, such as providing access to clean water. The orbital perspective would see lack of access to clean water as one symptom of larger issues related to education, unemployment, and government infrastructure. In a medical response to a natural disaster, medical personnel taking the worm's eye view would focus entirely on treating the patients right in front of them. The orbital perspective would treat those patients while also putting reconstruction projects in place to lessen the long-term risk that many more patients would end up "right in front of them." Poverty, inequality, corruption, and political instability multiply the negative effects of natural and human-caused disasters. Therefore, by reducing those social woes we greatly reduce the negative effects of disasters when they occur.

Living in the worm's eye view means putting the focus on the quarterly report, whereas the orbital perspective focuses on the twenty-year plan and beyond. The worm's eye view treats corporate social responsibility as a tool to prop up a brand's image, placing more emphasis on advertising social good than on contributing to that good. The orbital perspective makes the creation of social good a core business tenet throughout the entire organization, meaning the social, environmental, and ethical aspects of every business decision are carefully considered. This is done not only because the company has a responsibility to do so but also because in this age of transparency and openness, when the public has been empowered to scream from the digital mountaintops when corporations are not playing nice, not doing so will have a significant effect on the bottom line.

It's certainly fine to feel comfortable with and to care about the people in one's immediate area, those of the same religion or nationality, who speak the same language or come from the same tribe. But if we pull back to the orbital perspective, where all such differences blur into insignificance, we're able to see all the people and organizations involved in an issue, all the pieces of the puzzle. If we can't see everyone involved or all the pieces of the puzzle, then we need to step back further until we do. By "stepping back," I mean widening our sphere of influence and our prediction of the consequences of decisions to the largest geographic area and the longest time line possible. The trick is to step back far enough to create a frame for a problem in which everyone owns the problem and the solutions to that problem, enabling everyone to collaborate toward a solution. I call this frame of reference the orbital perspective, but as I have said, this frame does not require going to space. You don't have to be in orbit to participate in the orbital perspective.

In this view, directing all of your giving to a charity, where a significant portion of the money might go to overhead, may make less sense than investing some portion in a social business that operates like any other business, but with a primary and overarching goal to create some type of social and/or environmental good rather than to maximize profits, return on investment, or shareholder value. Now a much higher percentage of your money (potentially one hundred percent, if business operations are able to cover overhead) is going to create the intended good. Instead of creating a one-time, partial good, you're creating long-term, sustainable good. And if all goes well, the money invested will eventually be returned to you to do with as you wish, including reinvesting in another social business helping to solve another problem. In this model, then, "charity" as it is traditionally understood is reserved for those problems for which there is no feasible social business solution.

COURSE CORRECTIONS

There also is a temporal component to the orbital perspective. The worm's eye view focuses on a discrete moment, but the orbital perspective looks at every moment in the context of the whole life of an individual or group. It is an expansion of our awareness, from how our thoughts, actions, and words affect us and our immediate surroundings to how they affect the entire global society and the physical world. It is a movement from looking at the present moment as a discrete point on the time line of our lives to seeing it as a thread traveling from our complete past to our complete future. Each moment is intricately tied to the past, and it will forever alter the trajectory of the future.

When an individual or group is faced with a decision, it must first be noted that our radius of influence is significantly bigger than we think and that the duration of affect is significantly longer than we think—in fact it's infinite. The effects of any decision ripple out in all directions, changing multiple trajectories and gently (or not so gently) nudging the trajectory of our global society.

Science fiction is filled with stories of killer asteroids making their way to Earth to wipe out life on the planet. Usually, a team of Bruce Willis–like heroes mounts up and blows the asteroid out of the sky, saving the day. But actually, given enough warning, it would be a relatively simple, much less dramatic task to save Earth from such a catastrophic event. We could fly a small spacecraft out to meet the asteroid, and with the force equivalent to a feather, give it just a slight nudge, which would translate to a miss distance of hundreds or thousands of miles.[6]

Likewise, when we look at things from the orbital perspective we see that even a small change, early in the process, can translate to major changes in trajectory. Our past becomes a road map of where we've traveled and a forecast of where our present trajectory will take us in the future, and we can use the sum of our past successes and failures, good decisions and bad, to help us make decisions in the moment that would alter and adjust that trajectory.

VISION ZERO

Let's look at a more concrete example of the relation between the worm's eye view and the orbital perspective. Every day, in every city on the planet, people die from traffic accidents. The positive side of the worm's eye view would look at the details of a city's traffic safety record and try to fix things that lead to accidents—a campaign against drunk driving, more enforcement of traffic laws, better streets, or increasing automobile safety by designing cars that can better withstand a crash. The negative side of the worm's eye view would consider traffic fatalities to be an unavoidable and inevitable price we pay for our mobility. In fact, because serious traffic accidents are seen as inevitable, these incidents barely get mentioned in the news.

Pulling back to look at the situation from the orbital perspective, though, reveals that more than 1.2 million people are killed in traffic accidents each year—the equivalent of wiping out the entire population of a city the size of Dallas, Kampala, or Prague—and would view this situation as unacceptable. Such perspective would lead us to take a wider focus and to look for ways to incorporate information technology in the design of smart transportation systems that prevent cars from crashing in the first place.[7]

The orbital perspective also would take account of environmental and geopolitical issues, in addition to safety concerns. If by working together as a global community we can design an infrastructure that enables cars on the road to know the location of all other vehicles on the road, to know the road conditions, to employ sensors that react to any circumstance thousands of times faster than a human, we could, in effect, create a transportation system in which cars do not crash into each other or into anything else. Once we do that, we can take the armor out of vehicles and drastically reduce their weight and fuel consumption. We also could design innovative systems to reduce traffic, such as smart roadways, and smart traffic light and traffic flow sequencing. Reducing the amount of time people spend in traffic every day would drastically increase global productivity while significantly reducing fuel consumption, pollution and emissions, and reliance on fossil fuels.

This is not science fiction. We have the technology to do all this now. We simply need to take a bigger picture view and realize that no company, government, or consumer group will be able to effect an infrastructure change of this magnitude alone. We have to work together to achieve these higher flying goals.

THE ILLUSION OF DIFFERENCES

In 1966, Senator Robert F. Kennedy said, "Everywhere new technology and communications brings men and nations closer together, the concerns of one inevitably become the concerns of all. And our new closeness is stripping away the false masks, the illusion of differences which is at the root of injustice and hate and war. Only earthbound man still clings to the dark and poisoning superstition that his world is bounded by the nearest hill, his universe ends at river's shore, his common humanity is enclosed in the tight circle of those who share his town or his views and the color of his skin."[8]

Overcoming social woes starts, as Senator Kennedy put it, with an awareness of the "illusion of differences" and with pulling back until "the concerns of one inevitably become the concerns of all." After we take those first steps, we also need to employ elevated empathy. By acknowledging our unity and actively putting ourselves in the shoes of others with whom we share this world, we put ourselves on a true path to progress and advancement.

How far we progress and advance together is dictated by the framework we construct to delineate the possible from impossible. But this framework, if constructed within the orbital perspective, would embrace the mind-set that nothing is impossible. This framework would acknowledge that we live in a world in which our exponentially increasing technological advancements and interconnectivity are making the impossible possible on a daily basis. In the next chapter we will explore the frameworks we use to construct our view of the world and the implications of taking on a conviction that we do not have to accept the suffering and conflict on our planet as inescapable.

CHAPTER SIX
The Key Is "We"

✦ I recently attended a national summit on infrastructure at the Harvard Business School. One of the speakers, Senator Barbara Mikulski, told a story about how every month the women of the senate—Democrats and Republicans—meet for dinner. They meet without staff, without memos, and without the normal overhead that comes with political meetings in Washington. The dinners are designed to be all about relationships and listening. The idea is to create an atmosphere in which the senators do not judge each other. Senator Mikulski felt this to be a very powerful tool to break down barriers to cooperation and to provide a common ground that will serve as the relational foundation for working together. This is a great worm's eye view strategy, especially in this case, because the strategy has the power to overturn misconceptions, and in the extreme, to reduce the likelihood that one side will demonize the other.

When we pull back to the orbital perspective, we look not just at a group of people trying to make a difference but also at organizations, nongovernment organizations, and nations that can do the same. When we step back, the things that we share in common become more visible. If demonization, misunderstanding, and misconceptions can be overturned, common ground can be found, and that common ground can serve as a bridge toward the solutions to the challenges we all face.

As these senate dinners demonstrate, the first step to building collaboration with people and groups with whom you disagree, whose opinion differs from your own, is to find the common ground. It is important to use the easy things to prove that it's possible to work together, that it's possible to set aside the real and perceived differences that keep you apart.

The International Space Station program is a good example of the power of finding the common ground. The United States and Russia were Cold War competitors, both philosophically and doctrinally, but by doing something compelling, meaningful, and important together, the two nations were able to rise above their competitive inclinations and work together for a common cause.

As a fighter pilot during the Cold War, I trained for years to fight the Russians, but on April 4, 2011, I stood at the base of a Soyuz rocket named *Gagarin*, in honor of the fiftieth anniversary of Yuri Gagarin's first human spaceflight. As my two Russian crewmates and I waited to depart from the same launchpad Gagarin had used in 1961, I looked up and saw an American flag painted on the rocket—a compelling example of the transformative power of finding common ground.

SIMPLIFIED FRAMEWORK

Adopting the orbital perspective can have profound implications for our global society. But putting this perspective into practice requires us to first acknowledge the framework that we have constructed to view the world, which informs our opinions about events and situations. Our world contains suffering, injustice, crime, terrorism, natural disasters, war, and a long list of other woes. But the world also is characterized by love, charity, forgiveness, and interconnectedness. The complexity of the world, the enormity of its problems, and the sheer terror that can be inspired by world events are so great that there is a strong propensity to construct a simpler framework or worldview that attempts to incorporate all those factors into something more understandable and palatable. In this process of simplification, we tend to lump situations, people, groups, and nations into categories. It is much easier to label an entire group of people than to deal with the myriad complex factors that can lead people and groups to do things that we don't understand.

We saw how, in the early days of the U.S.–Russia space partnership, individuals on both sides had ingrained notions about their counterparts on the other side. Through their constructed, simplified frameworks, some people on both sides initially held on to an us-and-them mentality that led to misperceptions. Some of these misperceptions were confined to minor issues, such as not understanding aspects of each other's culture or collaborative style, but others were more serious, relating to each side's respect for the knowledge of the other, for instance, and these kinds of misunderstandings had great potential to derail the partnership before it ever really got off the ground. In some cases, simplified frameworks led to initial failures on both sides to empathize with the history and expectations of the other.

A simplified framework also can lead to a tendency to treat an event as a snapshot, or worse, as one in a series of snapshots that confirms the generalizations we've constructed about another, that disregards the history that led up to the situation and the possible global trajectory implied by that situation. Such a framework leads to hatred, racism, discrimination, violence, and war. It can lead us to ignore anything of merit on the other side, or even to attempt to discredit that merit. The building of a simplified framework leads to a simplified worldview, or nation-view, or community-view, or society-view and has led to sharp divisions in the global community and within many countries, including the United States. It replaces empathy and understanding with generalizations and egotism and represents an impediment to global cooperation—perhaps the main impediment.

Before making judgments on a situation—especially when decisions will be based on those judgments—we need to take the time to explore the worm's eye details of the situation while also pulling back to the big picture as it comes into focus in the context of the past, present, and future. We need to look at any situation in context and examine the various paths our reactions to the situation can put us on. Most importantly, we must try to put ourselves in the shoes of those we deem responsible for the situation. We must attempt to employ elevated empathy and seek a deep understanding of the situation before making critical decisions or sending out inflammatory tweets. Elevated empathy helps us to see the extreme opinions of others within the framework they are operating under.

Some of the extreme opinions on both sides during the early days of the U.S.–Russia partnership came about because the Russians had constructed a framework based on long-duration spaceflight experience aboard a series of space stations, whereas the Americans' framework was based on the space shuttle program's short-duration experience. Only after personal relationships were formed and people on both sides became willing to step outside of their comfort zones did each side begin to recognize the framework the other side was operating under and to see the merit of their different approaches.

PEACE IS WHEN MY BELLY IS FULL

In the mid-1980s, Emmanuel Jal was a young boy living in a small Sudanese village in the midst of fierce conflict. That was his worm's eye perspective, and the extent of his universe. He was barely seven years old when his mother was murdered by soldiers and he was conscripted as a child soldier into the Sudan People's Liberation Army. When the horrors became too much to bear, he and about four hundred other child soldiers, some of the Lost Boys of Sudan, escaped from their unit with some adult soldiers and began a three-month march toward Ethiopia. One by one, the boys died of starvation on their long march. Only sixteen survived.

Miraculously, Jal, barely eleven years old at the time, was one of those sixteen, and he ended up in a slum in Kenya. Jal was gifted with musical talent, and singing helped ease the pain of his life.

Today Jal is a successful international hip-hop artist and peace activist. He travels the globe sharing his music and his message of peace. In the process he recruits "peace soldiers," or, as he explained to me, those who use "their platform, talents, innovation, their entrepreneurial skills, or whatever they have to try to make Planet Earth better, to try to make a positive impact on humanity. Everybody's got something that they can do, and if you use that for doing good then you're a peace soldier."

As a child soldier, Jal was indoctrinated to kill as many Muslims and Arabs as he could, to avenge his family. "But I no longer hate Arabs, because my understanding has widened. I was young and blinded by rage, and now that I am older I know that if I ever fight again it will be for freedom, not hate."[1] The education Jal received in Kenya was instrumental in this transformation. "I learned we're all human beings. We can empathize, we can talk, we can share— this is what separates us from animals. We are all people. We eat, we bleed, we cry, we are the same."

Jal rose above his worm's eye view to see the bigger picture. He somehow knew that there was more to life, and ultimately he was able to rise above his circumstances, to put aside his preconceptions. He now uses the experience of his past, combined with his new perspective, to make a difference in the world. After successfully putting aside his hatred, Jal now believes "Peace is justice, equality, and freedom for all, and peace is when my belly is full. If you and I cannot eat, we are going to have an internal conflict. An empty belly does not understand law. If we are hungry we will not know law exists."

Jal pointed out the bitter truth that poverty and conflict are deeply intertwined. Years of development work can be wiped out in an instant when war erupts, and poverty itself is one of the biggest factors leading to war and terrorism. Poverty and conflict also influence every other problem and challenge facing our world, which is why we need to do everything in our power to bring peace and prosperity to the world.

I believe it is possible to live in a world without violence and armed conflict, where every person lives above the poverty line and the basic needs and rights of every person are met. I also understand how hard it is to fathom such a world of peace and prosperity, and one of the reasons for this difficulty is that we have been programmed to focus on the bad and dangerous. A small part of our brain called the amygdala serves as our danger detector. It searches our environment for things that could harm us. When the danger-searching amygdala meets the hyperconnected, information-overload world that we live in, it is no surprise that we pay more attention to bad news than good—and that there is so much bad news reported. When conflict is the big story, it's easy to miss the real story.

But the lopsided reporting of bad news gives us a distorted view of our world. Yes, we live in a very dangerous world with many problems and horrors, but the human spirit is underestimated against the sheer weight of the sea of negativity. Humans have a tremendous capacity to be resourceful, compassionate, ingenious, and forgiving, if we are able to set aside the framework we've constructed to view the world.

FORGIVENESS

Canadian Amanda Lindhout is roughly the same age as Emmanuel Jal, and although they come from seemingly different worlds, they both lived through catastrophic violence that made them warriors for human dignity. Lindhout is a stunning example of the power of setting aside what many would consider a natural worldview, in order to practice forgiveness.

On August 23, 2008, while working as a freelance journalist in Somalia, Lindhout and others were kidnapped by teenage insur-

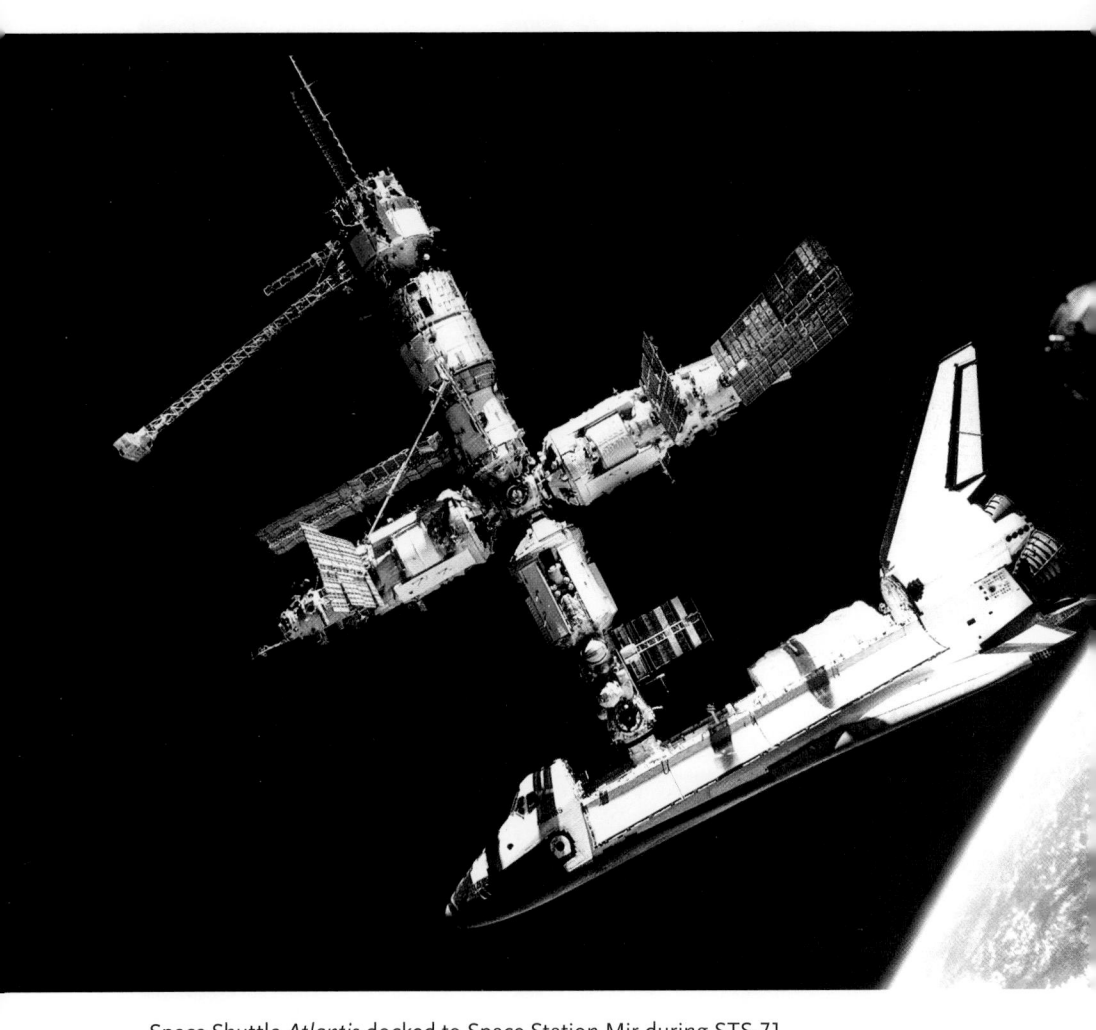

Space Shuttle *Atlantis* docked to Space Station Mir during STS-71.
This photo was taken from a temporarily undocked Soyuz spacecraft, while
Atlantis prepared to take Norm Thagard and his two Russian *Mir* crewmates
home, following the historic first joint U.S.–Russian long-duration mission.
(*Source:* Roscosmos)

Combined Manna Energy Foundation and Engineers Without Borders team installs solar panels on the roof of the Mugonero Hospital in rural Rwanda, 2006. *Left to right:* John Jannetto, Kiran Vinta, Niko Kalinic, Ron Garan, Evan Thomas, and Jean Pierre Habanabakiz. (*Source:* Engineers Without Borders-USA)

Speaking to children at L'Esperance orphanage, above the shores of Lake Kivu in rural western Rwanda, about the wonders of spaceflight, 2006. (*Source:* Garan)

Rocket-propelled grenade impact crater in a rural Rwandan church that was the site of a massacre during the 1994 Rwandan genocide. This violent scar, centered directly above the altar, symbolized hope that there is more out there than the two-dimensional violence of the parishioners' past. (*Source:* Garan)

Two months after returning to Earth, I gave a keynote address at the 2011 Global Social Business Summit in Vienna, Austria. Here, I am at the summit with Nobel Peace Laureate Muhammad Yunus. I presently serve as an ambassador and technical and scientific advisor to the social business movement and its founder, Professor Yunus. (*Source:* Dominique Dauster)

The crew of STS-124 heads to the "Astrovan" to start the journey to the launchpad, May 31, 2008. *Left to right:* Aki Hoshide (out of frame), Ron Garan, Greg Chamitoff, Mike Fossum, Ken Ham, Karen Nyberg, Mark Kelly. (*Source:* Eliot J. Schechter/Stringer)

An exciting day as I get suited up with survival gear in the White Room, prior to boarding Space Shuttle *Discovery* for launch, May 31, 2008. (*Source:* NASA)

Launch of Space Shuttle
Discovery, the Japanese Kibo
laboratory, and the crew
of STS-124, May 31, 2008.
(*Source:* NASA)

Although you can't see my feet in this picture, I like to think I had one foot in
day and one foot in night. Photo was taken on a spacewalk during the STS-124
mission, June 2008. (*Source:* NASA/Fossum)

Standing on top of the world. Here I am over Saudi Arabia on a spacewalk during the STS-124 mission, June 2008. Qatar, Bahrain, the Persian Gulf, and Iran are visible to the right. (*Source:* NASA/Fossum)

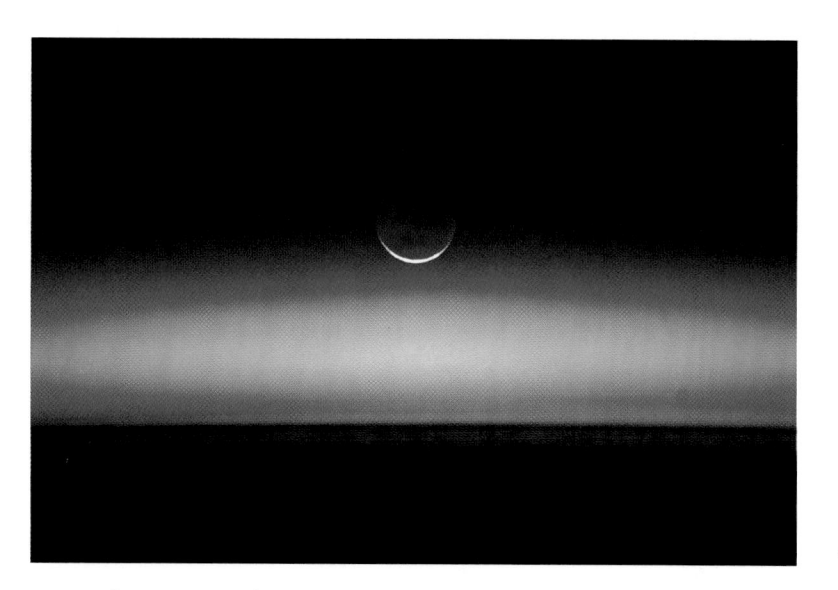

A tweet from space: "This is what the moon looked like 16 times today. #FromSpace We had simultaneous sunsets + moonsets," July 30, 2011. (*Source:* NASA/Garan)

I pretend to take a photo of Mike Fossum. Actually, I am taking a photo of myself and the ISS reflected in Mike's visor while on a spacewalk during the STS-124 mission, June 2008. (*Source:* NASA/Garan)

Chilean government and NASA officials in "Campo Esperanza" during the 2010 Chilean mine rescue. *Left to right:* JD Polk (NASA physician), Mike Duncan (NASA physician), Minister Mañalich (Minister of Health, Chile), two hosts from the Chilean Space Agency, Al Holland (NASA psychologist), Clinton Cragg (NASA engineer). (*Source:* Al Holland)

April 1, 2011, three days before launch. Standing with my Soyuz crewmates and our backups in front of the rocket that will carry our crew to space, we ask, "What kind of world do you want?" in English and Russian. *Left to right:* Dan Burbank, Ron Garan, Andrei Borisenko, Anton Shkaplerov, Sasha Samokutyayev, and Anatoli Ivanishin. (*Source:* NASA)

The Soyuz rocket that carried me and my Russian crewmates to space is brought by locomotive to the launchpad at dawn. Photo was taken by my son and photography coach, Jake. (Source: Jake Garan)

My Expedition 27/28 crewmates and I demonstrate our solidarity during a press conference on April 3, 2011, the day prior to launch. (*Source:* NASA)

April 4, 2011: A beautiful evening to launch to space. Our Soyuz rocket and TMA-21 spacecraft as we launch from Baikonur Cosmodrome. (*Source:* NASA)

July 12, 2011: I'm being "flown" toward Space Shuttle *Atlantis*, holding a failed pump module, with my feet clamped to the end of the robotic Canadarm2. Mike Fossum and I installed the module into *Atlantis*'s payload bay for return to Earth during the final space shuttle mission. (*Source:* NASA)

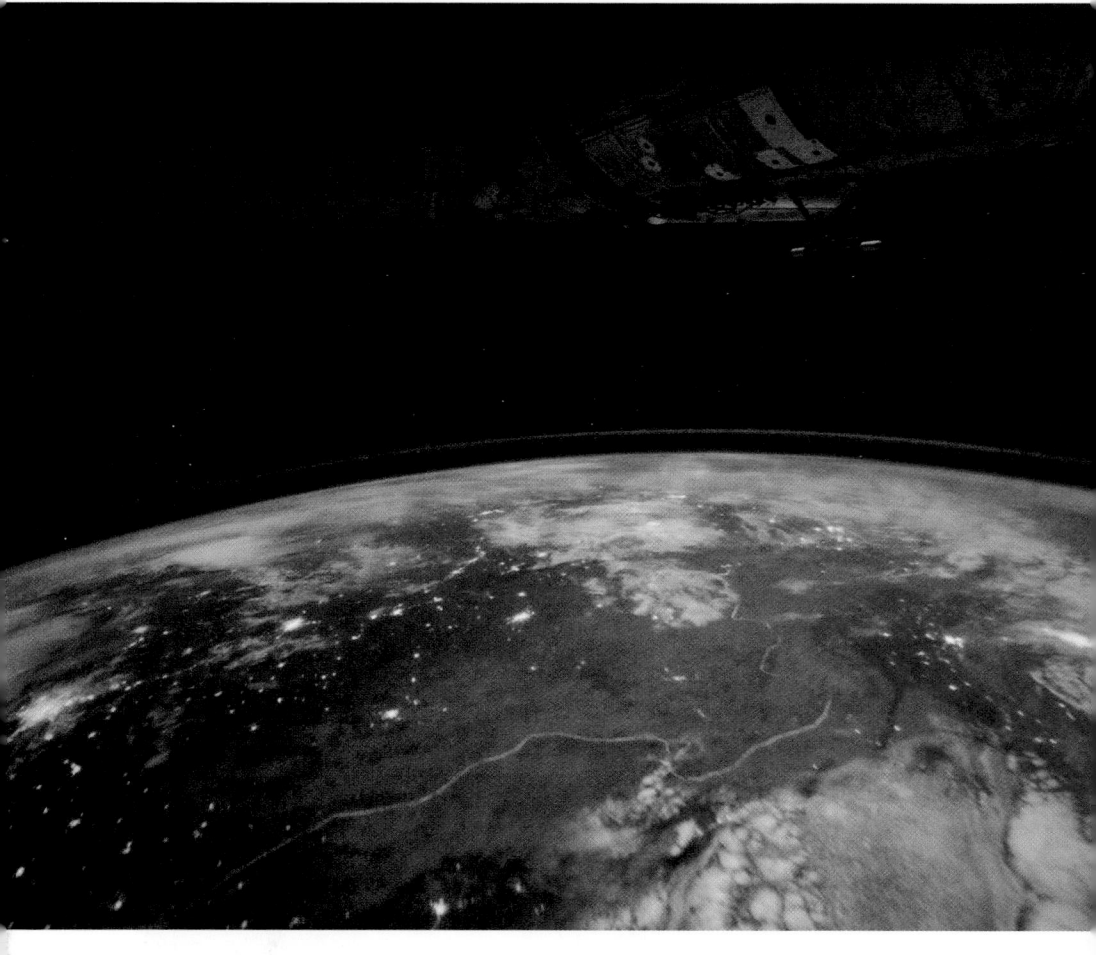

Illuminated human-made border between India and Pakistan. This scar on the otherwise beautiful landscape was a compelling call to focus on the need for global collaboration to overcome the world's problems. (*Source:* NASA/Garan)

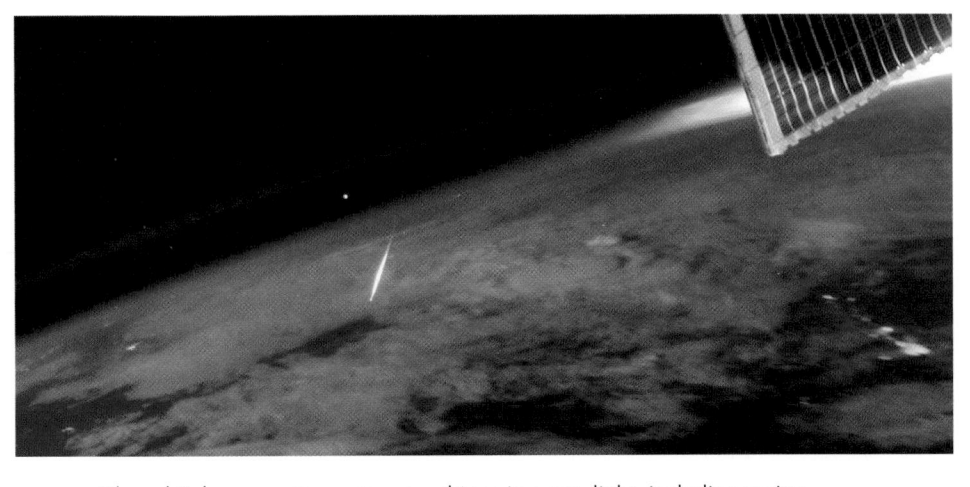

The orbital perspective puts many things in a new light, including seeing "shooting stars" *below* you! Luck was in my favor as I snapped this photo of a meteor during the Perseid meteor shower. At the time the photo was taken, the ISS was above the China–Mongolia border, August 12, 2011. (*Source:* NASA/Garan)

The dark blue of the Mediterranean Sea against the bright tans and reds of northern Africa, all framed by the curvature of Earth and its paper-thin atmosphere. Also framed within the picture is Libya, on August 24, 2011, the day that Tripoli fell during the Libyan Civil War. The violence and suffering, imperceptible from the orbital perspective, was all-encompassing when one zoomed in to the worm's eye details on the two-dimensional plane of the conflict. (*Source:* NASA/Garan)

A Home in the Sky. I felt like a high-altitude construction worker as I snapped this picture of the ISS and Space Shuttle *Atlantis* during the final mission of the shuttle program. Taken during the only spacewalk of the STS-135 mission, July 12, 2011. (*Source:* NASA/Garan)

An orbital perspective of Yemen, Ethiopia, and Somalia during the 2011 famine. Somewhere in this photo, Amanda Lindhout is leading the Convoy of Hope to help provide for victims of the drought in Somalia. (*Source:* NASA/Garan)

Space Shuttle *Atlantis* arrives at the ISS for the last time. I took this picture over the Bahamas, just prior to *Atlantis* executing the rotational pitch maneuver to enable documentation of the condition of its thermal tiles. (*Source:* NASA/Garan)

Mike Fossum catches me taking in the orbital perspective during the Expedition 27/28 mission, 2011. (*Source:* NASA/Fossum)

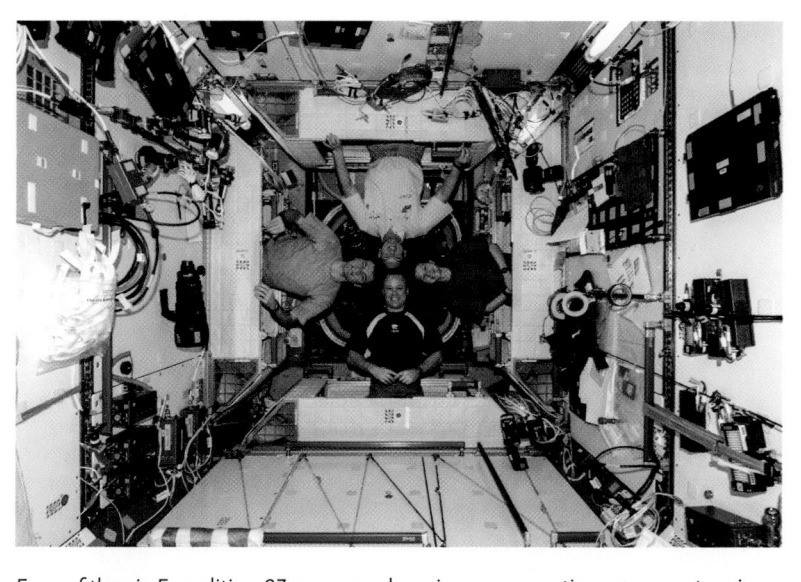

Four of the six Expedition 27 crewmembers in our respective crew quarters in Node 2 of the ISS. *Clockwise from top:* Russian cosmonaut Sasha Samokutyayev, NASA astronaut Cady Coleman, me, and Italian astronaut Paolo Nespoli. (*Source:* NASA/Borisenko)

After a fiery and violent ride through the atmosphere, my Russian crewmates and I descend toward "home" (Kazakhstan), under parachute in what's left of our Soyuz spacecraft. (*Source:* NASA)

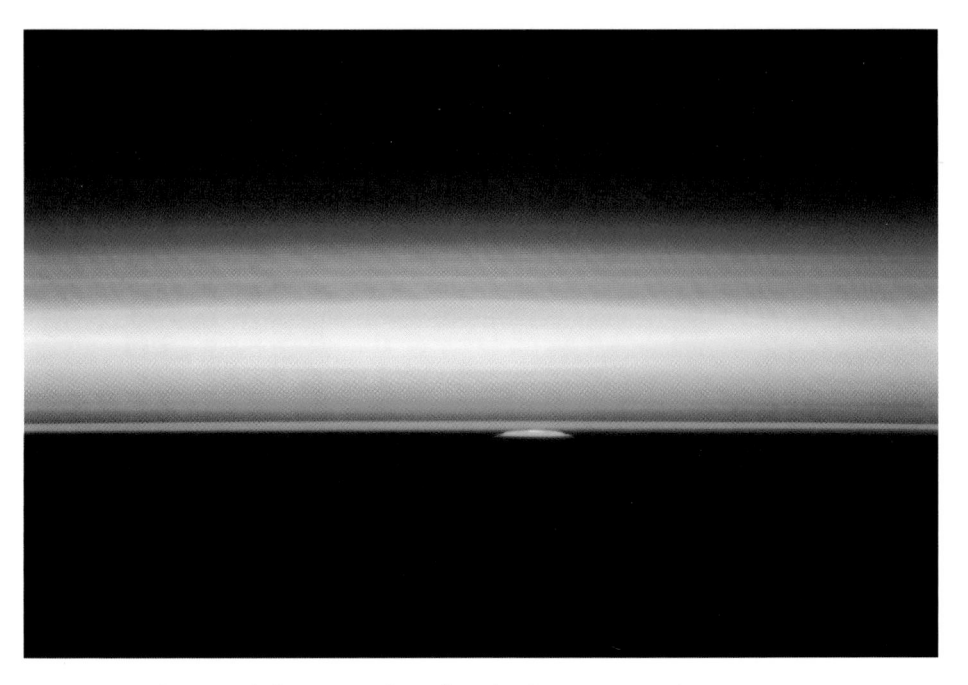

One of sixteen daily sunrises from the orbital perspective. This one was over Brazil, August 22, 2011. (*Source:* NASA/Garan)

Sunset over Corsica, Sardinia, and northern Italy, August 20, 2011. (*Source:* NASA/Garan)

gents from an Islamist fundamentalist group. After spending 460 days as a hostage, enduring unspeakable torture and abuse, Lindhout was ransomed and released, and the experience initially left her feeling hopeless. As she put it in her book, *A House in the Sky*, co-written with Sara Corbett, "I'd spent my life believing that people were, at heart, kind and good. This was what the world had shown me. But I couldn't find anything good about these boys, about any of my captors. If humans could be this monstrous, maybe I'd had everything wrong. If this was the world, I didn't want to live in it. That was the scariest and most disabling thought of all."[2]

Eventually, however, Lindhout was able to rise above her fear, hatred, and despair to see the bigger picture. She saw those that perpetrated violence against her as victims of violence themselves. They were the products of a war that had lasted more than twenty years and had orphaned thousands of children, just as another conflict had effectively orphaned Emmanuel Jal. "I choose to forgive the people who took my freedom from me and abused me, despite the fact that what they were doing was absolutely wrong."[3]

After being released, Lindhout fulfilled a promise to herself to find a way to help the women of Somalia. Six months after returning to Canada, Lindhout founded the nonprofit Global Enrichment Foundation to help support education in Somalia. "I'd spent so much time in captivity wondering about the boys who guarded me, specifically, whether they'd have been different—less entrenched in religious extremism and war—if they'd had more opportunities to go to school, and maybe more meaningfully, if they'd been raised in homes where their mothers and sisters had been able to attend school."[4]

This is an amazing example of rising above what most of us would consider a justified belief that the other side is purely evil and not even worth engaging, much less caring about. This is truly elevated empathy. Even more amazing is that Lindhout returned to Somalia. In response to the 2011 drought in East Africa, the Global Enrichment Foundation led an effort called Convoy for Hope to provide food aid. On August 4 of that year, Lindhout traveled back to Somalia for the first time since her ordeal, to lead a convoy to southern Somalia.

At the time, I was five months into my mission on board the ISS—an actual house in the sky—and I learned of Lindhout's incredible story on the uplinked news. Shortly after hearing about her story, as the ISS was over Somalia, I took some pictures of the area from space and sent out the pictures to highlight Lindhout's work from a slightly different perspective. I also used the IP phone on board the ISS to call Lindhout on her mobile phone during the convoy. I expressed my support and admiration for what she was doing, and she shared with me that she would never feel completely free until her sisters in Somalia are no longer suffering—a true demonstration that you don't have to be in orbit to have the orbital perspective.

Lindhout's Global Enrichment Foundation work in Somalia is an amazing example of global perspective and elevated empathy. Recognizing this, the Fragile Oasis community selected the Global Enrichment Foundation as the winner of the 2012 Fragile Oasis Prize for Humanitarian Service, for its work with Convoy for Hope. This global perspective and elevated empathy were made possible primarily because Lindhout did not allow herself to be imprisoned by her own limited framework. Instead, in a very heroic way, she was able to rise above her situation.

CO-LABORERS

After we have set aside the limited framework that may represent the largest impediment to mutual understanding, we have set the stage to form relationships and undertake global collaboration. Effective collaboration isn't easy. Creating and maintaining relationships takes work, collaboration can come at the expense of building new capabilities within existing organizations, and partnerships can incur high transaction costs. There's also much confusion about what collaboration means, and a tendency to reduce it to a buzzword. Nonetheless, the first step is to filter global collaboration through the lens of the orbital perspective, to begin to view our Earth as one community, as a complex, living system, rather than simply as an arena for self-interested competition.

The key to successful global collaboration is *we*, and the implied partnership is present in the word *collaboration* itself. Collaboration means to co-labor, to work together toward a common, overarching goal, placing the achievement of that goal far above the goals and desires of any individual team member. By committing to doing something grand and meaningful, the United States and Russia were able to overcome the legacies of the Cold War and produce a highly effective collaborative endeavor. The focus was no longer on racing to space for individual credit or glory. Instead, the two sides decided to work together for mutual benefit.

Successful collaboration doesn't require everyone in a group to have an equal vote or any solution to include all inputs. Instead, different individuals and groups bring together different pieces of the puzzle. Each member contributes something unique and adds value to the group. Solutions nobody thought were possible can come from places nobody ever heard of. As a result, this type of collaboration leads to solutions that are significantly better than any one member of the group could have created alone. Collaboration in this sense is characterized by diverse organizations systematically working together toward goals that benefit society as a whole.

ESTABLISH TRUST

As we've seen, one of the key ingredients of successful U.S.–Russian collaboration on the International Space Station program was trust, beginning with the building of strong personal relationships. The joint program in space didn't begin with agreements and plans to build an enormous orbiting research facility. Instead, to establish trust, the two sides agreed to intermediate steps that had less riding on them—sending astronauts and cosmonauts up in each other's vehicles, for instance, or undertaking a relatively simple docking maneuver—so that the failure of either side to live up to its agreements had a lower cost. If the partnership didn't work out, future collaborations could be easily shelved.

The Shuttle–*Mir* program required less time, resources, and effort than would be dedicated to building the ISS, and the cost of not living up to agreements rose exponentially with the ISS, which required significant investments of billions of dollars. Before undertaking the ISS, therefore, it was absolutely critical not only to have a proof of concept for the technical aspects of the partnership but also to have a demonstration of ongoing trust, so that all the international partners knew the others could be trusted to deliver what they promised.

One of the main things gained through established trust is a team's improved agility. Organizations that have built a foundation of trust will be able to handle unexpected or critical situations much faster and more effectively. On the other hand, teams that don't trust their team members to make decisions will attempt to spread the responsibility for decisions as widely as possible, and they will not act until they feel comfortable. The time it takes to move forward when a partner is making a decision is inversely proportional to the level of trust established in the partnership.

Therefore, another critical aspect of the maturing ISS partnership was that, as trust was gradually established, more and more authority and responsibility was given to the various ISS partners. After relationships are formed and trustworthiness and competence are demonstrated, an important component of true collaboration is this kind of shared control. It is important to recognize the value that partners bring to the table, and to step back and *let* them bring it to the table.

At the beginning of the ISS partnership, many at NASA did not appreciate what the Russians brought to the table, and vice versa. By working together incrementally, however, both sides came to see that *different* does not mean inferior. The two sides also learned that although team members may be unable to do something that you can do, they still can provide value. In fact, it is equally likely that they bring something that you can't, and thereby enrich the partnership deeply, as was the case among all the ISS partners.

Achieving this kind of trust, of course, requires team members or their leaders to share or even give up some control. True collaboration is not possible in an entirely authoritarian, command-and-control environment. Sometimes putting the overarching goals of the partnership above individual goals leads to working in ways you didn't plan and with which you may not feel comfortable, but the demonstrated ability to undertake the partnership anyway goes a long way toward solidifying collaboration on both sides.

Another great advantage of trusting a partner's judgment and ability is that it builds redundancy into the process, which may help prevent a single mishap from taking down the entire project. In designing spacecraft, engineers try to avoid situations in which the failure of a single component would cause loss of the ship, crew, or mission, and the best collaborative solutions likewise should not contain single point failures or depend solely on a particular individual or organization to ensure success. The best collaborations stand the test of time as individuals and organizations come and go.

LINES OF COMMUNICATION

Just as partners need to allow each other to provide expertise and act independently when necessary, a collaborative environment requires mechanisms that provide workers closest to the issues with a clear, open channel to communicate ideas, suggestions, and improvements to management. Those that have decision-making authority for a team or business must trust the creativity and intelligence of the members of their team. Likewise, management needs a clear line of communication to share its vision with those tasked with executing that vision. Obviously, the smaller and more centrally located a team is, the easier it is to build these types of relationships.

Leaders also must take a collaborative approach to working with their people. Leaders of the past were expected to have all the answers, whereas leaders today are expected to ask all the right questions—and listen to the answers. The traditional model of a hierarchical structure with a directive leader who acts as the gateway for information and decision making does not work effectively in our hyperconnected world.

To address the grand challenges facing our world, we will need to develop mechanisms to build these types of interactions among large, diverse, and dispersed organizations, and to overcome the challenges associated with collaboration among those who don't share the same language or cultural biases. In this type of environment it can sometimes be difficult to determine the right people to speak with or to figure out who possesses the necessary expertise to help with a specific problem. Social networks have the capability to bridge such expertise gaps, and technology has the capability to make processes considerably more efficient and to improve decision making by providing complete, concise, appropriate, and useful data and tools in a timely manner. People can find subject-matter experts within their own organization or, when needed, can enlist help from the global community. As we embrace this new technological world and virtual communities, however, we shouldn't lose sight of the need to build real personal relationships, and we should avoid the tendency to surround ourselves with only known quantities—things we already understand, or people to whom we are already connected.

SHARED CREDIT

Through all of this, it is important to give credit where credit is due. Nothing destroys trust and collaborations faster than team members taking credit for or ignoring the accomplishments of other team members. It took many years of working together before the ISS partners would publicly praise the efforts of the other partners. There has to be a willingness to overcome the mind-set that making other partners look bad, or lessening their contributions, makes us look better and our contribution seem bigger.

COLLABORATIVE NETWORKS

From the orbital perspective, collaboration is characterized not only by cooperating and working with those within your own company, nongovernment organization, nation, industry, or group but also by cooperating with everyone in your collaborative network. A collaborative network should include all individuals or groups that can bring value to a specific problem or situation, which may include anyone from the general public, employees, or management to competitors, clients, customers, or advocates. Someone who falls within one of these categories and is a co-laborer should feel some ownership of the task with which they're involved. This is particularly true when the overall success of a critical project is riding on a particular product or service. It is better to get some buy-in and establish a sense of ownership with the supplier of this product or service.

True collaboration that brings real value to a group usually means stepping outside of traditional organization and one's comfort zone to embrace new, innovative, and at times unconventional partnerships. If a potential collaboration looks too hard and messy, you're probably on the right track. If a particular collaborative partnership were easy, it probably would have been pursued long ago.

Truly effective collaboration involves motivating and inspiring individuals and groups outside of the formal control of your organization, who come from different cultural backgrounds and potentially speak different languages, to work together to achieve worthwhile common goals. This environment tends to lead to a somewhat flat organization that does not adhere well to the traditional, hierarchical command-and-control structure. As the need for innovation increases, the effectiveness of the traditional command-and-control structure can often decrease.

The following are some of the characteristics of effective collaboration, contrasted with ineffective collaborative techniques:

Effective Collaboration	Ineffective Collaboration
Team members resolve conflicts and differences at the relationship level and consider the overarching goals when determining the most effective way to assign limited resources or resolve competing goals.	Team members send conflicts up the chain of command and let managers settle the situation. They lobby their management to "rule" in their favor, and put their perspective over the overarching perspective (orbital perspective).
Team members are motivated primarily by the goals, needs, and objectives of the team.	Team members attempt to use the team to achieve personal goals.
Partners recognize the value that other partners bring to the table.	Partners believe that because an approach is different, it is inferior.
Co-laborers should feel some ownership of the task.	Team members only do things in their job description.
Partners in the collaboration are open to sharing control with their members.	Partners are territorial and engage in power struggles to gain control.
All partners give credit where credit is due.	Partners try to make themselves look better by making another partner look bad or by lessening a partner's contribution.
Team members are encouraged to look for better, more effective ways to do things.	Conformity is valued over innovation.

Effective Collaboration	Ineffective Collaboration
Team members cooperate with everyone in their collaborative network, including all individuals or groups that can bring value to a specific problem or situation.	Team members only cooperate and work with those within their own company, nongovernment organization, nation, industry, or group.
Willingness to step outside of the traditional organization	Stay within stovepipes
Emphasis on motivating and inspiring individuals and groups both inside and outside of the formal control of the organization	Emphasis on motivating those within the formal circle of control
Emphasis is on doing. Coordination occurs on as-needed basis	Meetings for meetings' sake
Big picture view	"How it affects us" view
Long-term view	Quarterly report view

COLLABORATION TO SOLVE GLOBAL PROBLEMS

To overcome the challenges facing our planet, we need to develop mechanisms to efficiently and effectively connect organizations and efforts to make life better for all its inhabitants. Such a platform should be open and accessible, with a continuum of resource and communication lifelines to pair challenges with solutions and bring them together.

We need an international team of people and organizations to work collaboratively to build backbone platforms that unify efforts across sectors to address and respond to humanitarian needs, quickly and efficiently. Optimally, the structure of such a platform would be completely open, allowing anyone to participate while also enabling the formation of subgroups customized to tackle specific problems or parts of problems, as needed. A completely open structure lowers barriers to entry and offers the possibility of attracting the largest number of problem solvers and the widest solution sets. The system should bring users that are closest to the problem into the system and ensure that everyone has a voice in the discussion.

Such a mechanism requires a collaborative community—a community of trust working together with a shared purpose—and a philosophy of contribution in which people are focused not solely on doing their assigned job well but also on making a good contribution to further the shared overarching goal. This framework would align people and organizations with shared purpose, but establishing, maintaining, and using this framework takes work on the part of individual team members. Effort must be expended to continually adapt and align efforts to synchronize with the overall effort.

NOTHING IS IMPOSSIBLE

Belief that we can actually collaborate to accomplish necessary global changes brings out an integral aspect of the orbital perspective: the belief that nothing is impossible. The exponentially increasing capability of technology is making the seemingly impossible possible on a daily basis, and there is probably no better example of this than the space program. For thousands of years, people thought it was impossible to fly to the moon and back, but human ingenuity and determination proved that it is possible. Likewise, many people believe that some of the biggest Earth-based problems facing humanity, such as poverty or long-standing conflicts, are impossible to solve. We tend to believe that things that have never been done before are impossible. But if we can land humans on the moon and return them to Earth safely, if nations can join together and build an enormous, incredibly capable research facility in orbit, we can, by working together, solve many if not all of the problems facing our planet.

To make full use of our rapidly increasing technological capabilities, however, we need to shift our mind-set and embrace the truth that nothing is impossible. This mind-set can empower us to attack the challenges and problems facing our world with vigor and confidence. We need to abandon a doom-and-gloom, defeatist's attitude and focus on collaborative efforts to solve our grand challenges. As the African proverb states, "If you want to go fast, go alone; if you want to go far, go together."

Great ideas, innovations, and breakthroughs can come from anywhere, and technological advancements have made it extremely economical to access those ideas and incorporate them into solutions. There is no reason for an organization or social enterprise to go it alone. In part 3, we will be looking at some of the many efforts to connect billions of minds and tap our collective global genius. We will further examine the barriers to global collaboration and explore ways to break down those barriers. And we will hear some stories about amazing people who are building bridges to collaboration and making great strides toward overcoming our world's biggest challenges.

PART III

LOOKING FORWARD

CHAPTER SEVEN
Camp Hope

✦ On a late starlit evening in September 2010, NASA psychologist Al Holland and his colleagues, exhausted and lost in their thoughts, were traveling in a van along dirt and gravel roads through the high-elevation Atacama Desert en route to the Chilean town of Copiapó. The group had just left the feverish activity of the dusty Campo Esperanza (Camp Hope), where hundreds of workers were laboring to free thirty-three miners trapped below 2,300 feet of hard rock. Holland had the hectic image of Campo Esperanza still fresh in his mind. The scene reminded him of something out of the movie *Close Encounters of the Third Kind*—a little valley sheltering huge lights that pointed at a single spot, with activity going on around the clock. That day, like every day since Holland and his group had arrived in Chile, had been insanely busy and filled with intense emotion.

As the van picked its way along the dirt roads, someone in the group noticed a couple of planets visible in the night sky and suggested they stop to take a look. As Holland and his colleagues piled out of the van and into the driest desert on Earth (also known as "the window to the universe," for its clear night skies), the contrast between the feverish intensity of Camp Hope and the grandeur of the new scene unfolding before them was palpable. "The Milky Way was just painted over the top of low black hills at night," Holland said. "It stretched in a great arc from a set of silhouetted hills behind us all the way across the sky to the hills in front of us. It was incredibly cool and quiet, dead quiet. This has been happening for millennia. For billions of years you've had this same coolness."

Compared to that cool, ageless, eternal, unchanging process, the fervent human activity in the nearby valley seemed like a flash in the pan. "It's like all the humans are focused on this tiny spot with such intensity, and in the big scheme of things [the mine] almost doesn't exist." In that moment, Holland stood in the continuum between the worm's eye view and the orbital perspective.

DOWN-TO-EARTH COOPERATION

It had been a little more than a month since a large part of the 121-year-old San José copper-gold mine complex shifted, resulting in the cave-in that trapped the thirty-three miners. Two days later, a second rock fall blocked ventilation shafts. Initial assumptions were that the miners were lost or would not be able to survive long enough to be rescued. Some experts put the probability of rescue at 1 percent.[1]

Two and a half weeks after the incident, on August 22, workers pulled the drill from the eighth exploratory borehole and found a note taped to the bit: "Estamos bien en el refugio, los 33 [We are well in the shelter, the thirty-three of us]." The discovery ignited a groundswell of international support to rescue the miners. Never had so many been trapped for so long and so deeply—in this case, all of the miners had found refuge in a five-hundred-square-foot space—and the operation to rescue them was unprecedented in scope, an incredible example of the power of elevated empathy and global collaboration.

In the early days of the rescue operation, it seemed unlikely that the miners would be rescued before Christmas, so the Chilean government was seeking advice on maintaining the physical and psychological health of the trapped miners for five to six months. NASA Administrator Charlie Bolden said, "The Chilean ambassador to the U.S. called me and asked for help. I said we'd be happy to help any way we can, but I'm not sure what we can do. The ambassador was adamant that if anyone could figure a way to get the miners out safely, NASA could." Holland, who has vast experience helping astronauts and our families deal with the isolation of long-duration spaceflight, was part of the small NASA contingent that contributed to the international response.

Four to five months of confinement is well within the range of situations that NASA astronauts face during long-duration spaceflight, but the conditions at the mine were much more difficult than usual NASA operations, and the miners and topside rescuers were not accustomed to operating within this type of confinement. Unlike astronauts, the miners had not been specially selected or trained, and there was no support system in place for their families. Nonetheless, NASA's long-duration spaceflight experience with the Shuttle–*Mir* and ISS programs did prove to be applicable to the rescue operation in this very different environment. Because the Chileans lacked NASA's experience in supporting people during long-duration confinement, they were all very open to suggestions.

The NASA team was one of many teams that arrived from around the world to consult with the Chilean government and the individuals who were doing the actual rescue work on the scene. Collaboratively, a comprehensive plan was developed to rescue the miners and to care for them and their families, both before and after the rescue. The rescue plan included the deployment of three large, international drilling rig teams, each employing different tactics to reach the miners. Multinational corporations from nearly every continent added their expertise and resources to the effort.

ESPRIT DE CORPS

When Holland arrived in Campo Esperanza, he met with his counterpart, Chilean psychologist Albert Iturra, who was from a small town not far from the mine. Iturra had worked with the miners before, and he knew them, their culture, and their families very well.

Iturra took Holland around to orient him to the site and introduce him to the various groups and people working there. They stopped near a small metal shack with a dusty, wooden floor, which housed an old desktop telephone that at the time was the only communications link between the miners and the topside rescue workers.

Iturra picked up the phone and began speaking with one of the miners. As he spoke, other support personnel came in, until eventually the shack was packed with people standing shoulder to shoulder. The speakerphone was engaged and a lively group conversation began. Toward the end, the topside people broke out into a cheer: "Chi! Chi! Chi!—le! le! le! Los mineros de Chile!" The miners began cheering from underground, and then people outside of the shack picked it up and began cheering.[2]

This modified soccer cheer, which frequently rolled through the camp, was indicative of the tremendous esprit de corps that developed among the operations community, which included not just the trapped miners and the rescue workers but also the several thousand family members camped in the hastily constructed tent city of Campo Esperanza. Other elements with a positive effect on the operation included the government's appropriate infrastructure response, including a small school; provisions for food, water, health care, and sanitation; and even entertainers to help keep spirits high.

Holland mentioned some other, similar measures that were used to keep morale high. For instance, when people needed fresh shirts, they were given shirts with logos on them, which fostered a sense of teamwork, community, and mission, just as in official NASA group photos all team members wear a mission patch. "You actually leverage all those opportunities to solidify the team and solidify the individual within that team."[3]

BIG PICTURE PERSPECTIVE

When Holland arrived at Camp Hope, there was a feverish intensity to the operation, a sense of desperation. People felt, understandably, that time was of the essence, that the miners needed to be gotten out of the mine immediately. Families were camped at the site, leaving their homes unattended, and rescue personnel weren't taking breaks or rotating out, increasing the chance of burnout. And, Holland said, "The miners themselves were desperate to get out and were seeking ways to get out which weren't effective, weren't working, and they were thinking in short-term, survival mode." One of Holland's first goals, therefore, was to realign expectations and establish a realistic time line. "I wanted to slow. The whole. Thinking process. Down," he said. "I wanted to put everyone back in their chair and say, 'We're here until Christmas.'"

The key was to shift everyone over to thinking of the effort as a marathon rather than a sprint. Soon, the Chilean minister of health picked up the marathon analogy and began speaking to the workers, families, and press using those terms, which helped everyone refocus their expectations and enabled the operation to take a much more appropriate pace. Establishing an end point (at first, five to six months) enabled rescue workers, families, and the trapped miners to focus on the light at the end of the tunnel and proceed systematically toward it. In addition, the challenging nature of the rescue operation itself helped to focus the efforts of all involved. "If it's time-limited, and particularly if your constraints are tight and it's a challenge to do," Holland said, "people tend to come together very well, and they can sustain a focused effort and can have collaboration for a time-limited period. But they have to know it's going to end, because otherwise they would exhaust themselves."[4]

Holland knew that being successful meant everyone had to take a big picture perspective. "You can't just focus on the miners. You have to realize that individuals live in social systems, and one way to change an individual's health, well-being, behavior, and performance is to intervene in the other systems or the other people that influence that individual. In this case, there were three key groups: there were the families of the miners, the miners themselves, and the topside personnel. You have these three key groups, and they're interlocking, they affect one another." All three of these groups needed to shift their mind-set and expectations.[5]

HUMILITY

From the outset, the Chilean government had the humility to acknowledge that it did not have all the answers. They took in information from outside sources, such as NASA, and made use of the advice that made sense to them. As we've seen, this is an important element in any collaboration. In addition to the shift from sprint to marathon thinking, Holland was able to offer a number of other techniques that derived directly from his knowledge about work with teams in space, space simulations, submarines, polar ice stations, and other environments—experience that the Chileans may not have had. For instance, it was important to create a community underground and then link that community with the community on the surface. The miners needed to systematize their circadian rhythms, and they needed activities they could participate in to work toward extracting themselves.

"There was a wide range of things from our experience that we could simply hand over to them," Holland said. "In fact, I took some of the training stuff that we had on repatriation after long-duration spaceflight and I had it translated into Spanish and just handed it over to them and they used it. . . . They were open to anything. So it was a very rapid impact."[6]

Much of the ultimate success of the operation is directly attributable to the Chileans' willingness to be very forthcoming with their needs and the gaps in their experience and to be open to ideas from the outside. There have been disasters of similar or greater scale where countries tried to go it alone, leading to less than desirable outcomes. The Chilean attitude was partly a result of humility, but it was also because the stakes were very high and the outcome was so uncertain. The lives of the miners were hanging in the balance. The Chilean government also had been criticized for a poor disaster response to a devastating earthquake six months earlier and did not wish to repeat this misstep. None of the members of the leadership team, including the Chilean ministers, felt as though they had all the answers; thus, they were more open to trying new things, considering new ideas, undertaking more open collaboration, and moving away from a traditional, hierarchical command-and-control system.

BENEFITS OF A SHORT COMMAND CHAIN

The relatively low level of bureaucracy, which Holland characterized as a short command chain, was another critical component of the success of the operation. NASA team members and mining experts had direct access to the ministers appointed to head the operation. "That's rare, for a technical guy to provide immediate information that is acted upon and you can see the feedback loop immediately. . . . It was remarkable to see a minister-level person in a Quonset hut, in field gear, with scuffed up boots on, in front of a white board, talking turkey. The nonbureaucratic aspect of it was refreshing." If anything was needed or a critical decision needed to be made, one of the ministers could make the decision himself, or call the president, and it got done. This structure saved precious time, and because the teams were racing against the clock, saving time meant saving lives. "It was really different than your usual day-to-day work setting where there's a lot of bureaucracy in a large government organization. It wasn't that. It was a field intervention, which moves very rapidly, so they were set up for that and they were open to that, and that's why it was rapid and effective."[7]

However, this refreshing lack of bureaucracy became apparent only after the NASA team arrived on-site. Before the group was able to make a meaningful contribution, a considerable amount of bureaucracy needed to be overcome. For instance, the team's initial recommendations document wound its way through NASA headquarters and the State Department, down to Chile and the minister of health, but as the team discovered later, it never made it to the specialists at the mine. "To us, waiting in Houston," Holland said, "the official permissions and official written requests seemed to grind on forever. There's a lot of diplomacy that needs to occur at the higher levels, but at the working level, you know what to do, you're ready to go do it."[8]

The delays were particularly frustrating because time was so critical. The miners were trapped in a small main area with very high heat and humidity; before they were discovered, they had very little food or potable water. Psychologically, Holland knew that things could get really bad under such conditions. He knew of fist-fights breaking out during long-duration confinement. There also might be other social problems, health problems, even suicides. If the miners' capabilities and coping abilities were not sufficient to the task, they would struggle greatly. The pressing need at the mine, as Holland saw it, was to raise the miners' capabilities and coping abilities to a level where these things would not develop. Every day that went by made this more challenging.[9]

COMMON CAUSE

The mine rescue had a very straightforward overarching goal: to bring all thirty-three miners out alive and in the best shape possible. It was a fairly easy thing to rally around, and this common cause led to cohesion of action, or as Holland called it, going into mission mode. And to some extent, the whole world rallied around the goal. Data, resources, suggestions, and other contributions flowed in from universities, institutional researchers, private companies, government agencies, and the public. The Lighting Science Group Corporation in Satellite Beach Florida contributed special lights that would fit down through the borehole and had the correct blue frequencies required to help entrain circadian rhythms and get the miners' sleep–wake cycles settled. Another company came up with a folding screen that fit through the borehole; this was sent down, along with a camera and small projector, enabling video teleconferences with the miners' families at the surface. And the Chilean government put out an innovation challenge of sorts, seeking designs for cots that could be packed into a four-inch tube. The response to this was enormous, and within days the team had several different prototypes and started sending cots down to the men.

Sending down cots was more than just a way to ensure the miners were as comfortable as possible; it probably saved some lives. One of the results of sleeping on hot rock is muscle breakdown, and the products of that breakdown have a detrimental effect on the kidneys. The rescue team was monitoring the men's health as carefully as possible, and when the topside team sent down urine test strips, sixteen of the thirty-three miners tested positive for high myoglobin, which indicated muscle breakdown. These men were in the early stages of kidney failure.[10] By getting the miners cots (and by increasing their hydration), the medical team was able to avoid this dangerous outcome.

The rescue itself presented another challenge. There were many concerns related to physically extracting that number of people through a 26-inch-diameter hole traveling through so much rock, including ensuring and monitoring the men's health, guaranteeing oxygen supply, preventing panic, and preparing for an emergency such as a stuck extraction device. To address these and many other questions, the Chilean government asked NASA to design specifications for the extraction device. Team member Dr. James Polk, chief of NASA's Space Medicine Division, said, "So, much like when we build a space capsule, we built the requirements. What are the physiological requirements, the medical requirements, the cage requirements? . . . We had pages and pages of requirements, written in typical NASA fashion."[11]

The Chileans took NASA's requirements and sent them out to industry, academia, and other places, asking that people send in their best ideas for an extraction device. In response, engineers, students, industrial and military organizations, and creative people of all sorts built prototypes or conceptual models, and in short order there were several working prototypes on-site. "The Chilean Navy actually had the best design, and so that's what they went with . . . based on NASA requirements. The Chilean Navy did tweak some things and made some upgrades and had some interesting ideas as well that parlayed a lot of our different requirements. It worked very well."[12]

In all cases, people rallied around the common cause. The overarching goal united people from around the world into a community with a shared purpose and motivated diverse groups to work together systematically and selflessly toward the rescue of the men. In short, the effort represented collaboration from an orbital perspective.

SPLASH UP

On October 13, 2010, shortly before 10:00 p.m., the last miner to be rescued, shift foreman Luis Urzúa, arrived at the surface of the mine. After sixty-nine days trapped deep under the ground, it had taken nearly twenty-four hours to bring every man safely to the surface. As the winning capsule design, which looks like a cross between an industrial elevator car, a torpedo, and a space capsule, was winched to the surface, joy exploded throughout the camp. Confetti and balloons flew through the air, champagne flowed, and cameras flashed everywhere. The camp erupted into "Chi! Chi! Chi!—le! le! le! Los mineros de Chile!" Urzúa emerged from the capsule draped in a Chilean flag, while rescue workers poured champagne all over each other, danced, and screamed with joy.

After embracing his family members, Urzúa embraced President Piñera. A BBC news commentator covering the scene reported, "The boss below ground hugging the boss above ground. They come from completely different backgrounds. One is a billionaire born into great wealth, the other a simple miner . . . and here they are, united in the resolution of this long, drawn-out ordeal and crisis."[13] With tears in his eyes, Piñera told Urzúa, "You're not the same after this and neither are we. We will never forget this."[14] The president led everyone in singing the Chilean national anthem, all with their mining helmets over their hearts, ending with a last "Chi! Chi! Chi!—le! le! le!" Appropriately, the commander of the ISS, Doug Wheelock, had earlier radioed a message on behalf of the ISS crew, congratulating "the heroes both above and below the ground."[15]

FOCUSED COLLABORATION

The outpouring of support for the Chilean miners and their rescue was a truly focused collaboration. Al Holland noted that the effort "gave me a lot of hope in humanity." From a technical point of view, the rescue also demonstrated the power of interdisciplinary collaboration. Two seemingly unrelated environments—outer space and subterranean space—meshed well into a series of workable solutions. As James Polk stated, "A lot of the lessons from spaceflight . . . came into play—the psychological aspects that we've learned on ISS probably more so than we even anticipated in our first conversation or two. And, fortunately for us, it was a very positive outcome, with the thirty-three people surviving."[16] By engaging with NASA, the Chileans were able to access the entire space community. "Because of the networking that we have from the International Space Station," Polk continued, "if I needed something, I could reach out and touch the Japanese flight surgeon, the ESA [European Space Agency] flight surgeon, the Canadian flight surgeon, the Russian flight surgeon. We are a global network."

Technology also played a big role in the operation and enabled much of the international collaboration, as well as communication with the miners underground. Social media, e-mail, Skype, and other communications capability enabled quick organization and information transfer between the two hundred or so people on-site and the rest of the world. And the entire world stepped up with ideas, designs, and support.

Another key factor in the collaboration was the operation leaders' insistence on open, honest, two-way dialogue, and their leadership by example. For instance, Mining Minister Laurence Golborne made a point to communicate not only the effort's successes but also its failures, whether he was speaking to the trapped miners, rescue workers, families, or reporters. "The decision of transparency was a conscious decision made early on," he explained. "There were too many people; we could not hide anything. If we did, we would lose their confidence."[17]

Openness to new ideas ebbs and flows within organizations, even within NASA. The leaders of the rescue operation were open to new ideas and dialogue because there was a high level of uncertainty, the stakes were high, and their experience level was low in this situation. Uncertainty and risk go a long way toward motivating collaboration and openness in large organizations, said Holland. "Then they become more collaborative, they become more willing to experiment, they become more selfless because they're afraid of the downside."

The Chilean rescue represented a global effort to solve a worm's eye problem, and in this case everything went right. In the next two chapters we will be looking at some things that can hamper such unified efforts, particularly in relation to the work of development organizations. We also will look at some efforts that are succeeding and will examine the value of community participation on a global scale.

CHAPTER EIGHT

Arrested Development

✦ The rescue of the Chilean miners, like the ISS collaboration and the U.S.–Russia partnership before it, was a great success and a great example of the orbital perspective in action. Despite our technology and resources, however, we still have many crucial problems, and in general we seem to be unable or unwilling to collaborate on a global scale. This is true even in fields such as development work, where we would expect to find—and often do find—a greater percentage of people who understand the orbital perspective and who practice elevated empathy.

INCENTIVIZING CHANGE

In the fall of 2004, I met a twenty-one-year-old student intern from the University of Colorado named Evan Thomas, an aerospace engineering undergraduate working at NASA's Johnson Space Center. I was interested in development work, and Evan had already amassed a fairly impressive amount of it. He had traveled the globe to such places as Cuba, Nepal, and Vietnam and was the founder of the Johnson Space Center chapter of Engineers Without Borders (EWB-JSC).

At the time of our meeting, Evan was just getting back from a development trip to Rwanda, which he had initiated and led. Evan and several other EWB-JSC volunteers described the projects they were working on, and I described to them the nonprofit I had founded, the Manna Energy Foundation (MEF). Initially, MEF was founded simply to promote renewable energy, because the more we learned about renewable energy, the more we saw the connection between the provision of energy and the alleviation of poverty. But from the outset the other founding members and I wanted the effort to be financially self-sustaining.[1] At the time I had never heard of social enterprise or social business. I knew only that if we could develop a business model that would support the humanitarian objectives of the organization, we would be able to do long-term strategic planning and not have to go from fund-raiser to fund-raiser.

The Manna Energy Foundation has since evolved to become a social enterprise incubator with the goal of harnessing the power of business to provide financially sustainable mechanisms to solve social and environmental problems. Since taking on that purpose, MEF has launched Geeks Without Frontiers, spearheaded by serial entrepreneur Michael Potter, which is providing broadband mesh Wi-Fi solutions to entrepreneurs in developing countries, and Manna Energy Ltd., which was launched to attempt to fund water purification projects through a new and completely untested financing model.

In January 2006, I went to Rwanda as part of a joint EWB-MEF development trip. Soon after, Evan joined the efforts at Manna, while initially continuing his duties as president of EWB-JSC. Before joining Manna, Evan had been experiencing some frustration with the way the projects he was running were working. He saw a certain level of futility in putting a great deal of effort into raising twenty to thirty thousand dollars to do a development project, only to see the project later fail when it couldn't raise additional money to keep going.

Evan was looking for a way to scale the impact he wanted to make and to do it in a way that would keep the long-term view in focus. He was looking to create positive interventions that lasted, that used appropriate technology, and that both fit into the big picture, long-term landscape of the region and incorporated the worm's eye details on the ground in the local community. In effect, Evan was looking to make an impact from an orbital perspective. Since the Manna Energy Foundation was also looking for a similar financially sustainable model, it made sense to join forces.

In 2007, we focused in on carbon credits, a United Nations system created under the Kyoto Protocol. We created Manna Energy Ltd. to combine our expertise in Rwanda with a carbon credits-for-drinking water concept we developed. Specifically, our desire was to find a way to fund projects that could provide access to clean water for the nearly one billion people around the world who do not have access to it. Over the course of the next six years, Evan led the Manna Energy Ltd. team that developed the work in Rwanda into the first United Nations program to be eligible for carbon credits for the purification of water. In 2011 this model led to a collaboration with the multinational textile and social enterprise company, Vestergaard Frandsen, providing nearly 900,000 family-size water purification filters and affecting approximately 4.5 million people in rural Kenya. It also led to an effort, which began in 2013, to provide water filters to the poorest 25 percent of the entire country of Rwanda—nearly 3 million people.

All of these efforts were and are predominantly funded through the generation of carbon credits, which are issued and then sold to buyers only after impact is demonstrated, as opposed to promised. We also are trying to encourage a focus on maintaining existing projects rather than starting new ones. Therefore, the implementers of these carbon credit–funded projects do not get paid for putting them in; they get paid for demonstrated use, based on the number of clean liters of water that pass through the filters. This creates a tremendous financial incentive to keep the filters working, to conduct effective training programs, and to implement processes to replace the filters when they've reached the end of their useful life. The high level of continual rigor that goes into the documentation required to verify the carbon credits leads to a significant boost in the accountability of the program.

The collaboration in Kenya led to the first carbon credits from drinking water issued in the world. In 2013, DelAgua, a UK water testing company, purchased the Rwandan assets of Manna Energy Ltd. and is now investing up to $60 million in Rwanda. The Tubeho Neza ("Live Well") project in Rwanda is being conducted by the newly formed DelAgua Health. Evan Thomas is leading this effort in Rwanda under the DelAgua banner.

Working on these and other projects, we have developed a close relationship with the development and aid communities both inside and outside of the government sector. In addition to providing access to clean water, we also have gained by having to confront and overcome unforeseen challenges, which have provided perspective on the development community as a whole.

IS THE SYSTEM BROKEN?

The majority of people involved in global health, development, and poverty reduction work are good people who want to have a positive influence on the world, and they work really hard to make their projects work. However, despite the billions upon billions of aid dollars that have flowed from developed to developing nations over the past decades, and the hard work and sacrifice of millions of good people across the globe, our global society still faces a mountain of huge challenges. We have made some great strides toward overcoming such challenges, but if we compare the progress we've made with the progress that *could have been* made if we had applied more rigor, discipline, and cooperation, and if we had taken a long-term, big picture perspective rather than a short-term view, it seems fair to ask whether the system itself is broken.

It's easy to look at this disconnect and to blame something like government corruption in developing countries, but we can't simply blame everything on corruption or on any other single factor. If we maintain the big picture perspective while zooming in to the worm's eye details, we see that the very systems put in place to tackle the problems facing humanity are, in large part, a problem in themselves. Many factors make our global development organizations less effective than they should be, including a lack of effective feedback loops, a failure to make a demonstrable impact, a focus on short-term thinking, an overemphasis on "innovation" at the expense of proven approaches, an organizational structure that features a lack of experienced midlevel management, bureaucratic inertia, and destructive competition among organizations that ought to be allies. Many of these issues also affect governments, and in all cases, they can affect progress toward solving global issues.

LACK OF FEEDBACK AND DATA

One big disconnect in development work is a lack of effective feedback: evidence-based, quantitative data that decision makers can use to target programming and resources. There are experts in impact evaluation, but for the most part, international projects rely on measuring efficacy through community surveys. Sadly, this feedback mechanism is both expensive and notoriously inaccurate. Interviewees tend to say what they think their questioners want to hear, which can result in assessments that are both inaccurate and paint a far rosier picture than the reality.[2] Some of these issues were experienced during the water filter rollout in Kenya, and after learning from that experience, DelAgua's rollout in Rwanda will incorporate quantitative sensors on a statistically significant number of filters, which will provide unbiased "truth" data.

Funders need better ways to evaluate the efficacy of the programs they have funded. Unfortunately, even after solving the problem of data bias, many organizations remain reluctant to share data that would show the true efficacy of their projects. Vic Gundotra, who eventually went on to become an executive at Google, found this out in 2007, when he tried to tackle a problem that had concerned him for years: how to enable better collaboration among organizations. Gundotra realized that, even among humanitarian organizations that shared a common goal, there was a lack of cooperation and collaboration that was severely hampering efforts to solve our world's greatest challenges. With his strong background in computer science, Gundotra devised technical solutions to enable collaboration among organizations. What he soon learned, however, was that the cultural barriers to collaboration were much harder to overcome than the technical barriers. The heads of nongovernment and other humanitarian organizations seemed very eager to participate in the data sharing programs that would enable collaboration, but when it came to actually opening up their own data stores and sharing information, the process ground to a halt. Gundotra found that humanitarian organizations, for the most part, were unwilling to share the level of data necessary to enable true collaboration.

There's a big difference, for instance, between releasing data on the number of hand water pumps an organization has installed and disclosing the number of hand pumps installed *that are still functioning.* Organizations are usually very willing to share data on the former and are not so open with the latter, because the number of projects that remain effective over a long period of time is relatively small, and also because there is a lack of real quantitative data defining what's actually working and what's not. Reports that show that half of all the money spent in the previous ten years was wasted are less likely to see the light of day than reports that say that so much money was spent and so many hand pumps were installed.

Likewise, there is a tendency to take pictures of a new system going in and to note how many people have clean water now, but not as much effort is put into going back and making sure things are still working. There can be a fear that if the actual long-term success of a project is measured too vigorously, and the real data becomes too publicly known and studied, perceptions of success will fade. In a donor-based environment, this could be the kiss of death. This can create perverse incentives that lead organizations big and small, public and private, to focus their reporting on superficial success and to avoid releasing complete data. One solution to this is to equip development projects with inexpensive sensors that provide true quantitative data, and then to tie an organization's continued funding to the sharing of that data—that is, to make open data sharing the norm among funding organizations.

Looking at a successful initiative such as the Chilean mine rescue, for instance, we see that transparency and openness—even when the news was not always good—eventually led to a positive outcome. The motivation for this openness and transparency in that case was that the stakes were high and lives were hanging in the balance. If we take an orbital perspective and a long-term view, we see that the same is true for our global community. The stakes are in fact very high, probably higher than we realize, and accurate information, transparency, and openness are key to addressing our global challenges.

FAKE IT TILL YOU MAKE IT

Because there is so little evidence, feedback, or accountability in the development arena, it becomes ridiculously easy to create an unchallenged reputation for excellence. By talking the talk and being on the conference circuit, an organization can be very successful without evidence of superiority or proof of having made a positive difference.

Many initiatives begin with a "fake it till you make it" attitude. New organizations get a splashy website and some celebrity endorsements, throw big fund-raising bashes, and get on the speaking circuit. As a society we tend to expect—even admire— this behavior. If you fake it and then actually succeed, the rewards are great. Everyone admires your gutsiness, persistence, and ambition. But there is a subtle difference between faking it till you make it and "making it by faking it," which crosses over into duplicity and deception. Unfortunately, in global development efforts, many organizations can have all the hallmarks of success without true, demonstrated impact.

SHORT-TERM THINKING

Most development money comes from government aid agencies that operate under a fiscal year cycle. They have to spend money according to dictated legislative priorities that can change as often as every twelve months. Sometimes they have no money to spend, and other times they have to spend an entire year's budget in a few months. This short-term focus can lead to situations in which government program managers are actually incentivized to cancel the most cost-effective, high-impact programs, if these programs aren't spending money fast enough. Funding less-efficient projects that cost a lot of money may even end up boosting the credentials of a program manager, because the money is being spent.

In the case of the government, relatively frequent changes in leadership tend to lead to changes in policy objectives. A new government may want to shovel money out the door to demonstrate immediate change and impact. In the process, new projects tend to win out over established projects.

This situation generally is the result of funder actions and priorities. People naturally want to see tangible, immediate results. To some private or institutional donors, donating money for five years of pump maintenance is not as attractive as getting a photo of a new hand pump in place in an impoverished village. From the orbital perspective, the pump maintenance would have a much greater and longer-lasting impact, but it may be less satisfying from the personal perspective. Unfortunately, it is the implementer and those they serve who suffer. Many implementers would be overjoyed to be given the flexibility to dive into a long-term engagement, but the system that the funders work under favors short-term successes.

OVEREMPHASIS ON THE NEW AND SHINY

Related to this short-term approach is the excitement over innova-tion, and a system that rewards spending money on new programs or on things that can be defined as innovative, both of which can negatively affect development. I am a strong proponent of innova-tion, but that excitement needs to be balanced. There is a risk in putting too much emphasis on the new and shiny.

Unfortunately, there presently is not a lot of reward for staying the course and continuing the hard slog to do things that actually work. Nobody gets to give a major conference talk about something that has worked for twenty years and is now in maintenance mode. As we've seen, an organization usually is not rewarded for main-taining an existing water pump; it is rewarded for putting in a new water pump. And again, this perverse incentive negatively affects development efforts.

So, for instance, thousands of water pumps are installed every year by implementers, but Evan Thomas and colleagues from Oxford University have come up with some surprising estimates about just how effectively the money for these pumps has been spent. "Based on a single year's expenditure to install new pumps, we found that if we diverted the money to put in the number of pumps normally installed in one year, we could double the main-tenance budget on existing pumps for the next twenty years."

From a cost-effectiveness perspective, therefore, it would be a lot smarter for a grantor to provide $1 million to maintain five hun-dred water pumps for the next twenty years than to install more new pumps without a maintenance budget attached. But that's not usually the way funders think. Most rewards are given for doing new things or undertaking new construction—rarely does anyone get credit for maintenance. Instead, maintenance becomes the community's responsibility.

If a humanitarian organization exists within a donor-based structure that disproportionately favors the new and shiny over the tried and true, the organization is incentivized to put undue emphasis on new projects at the expense of maintaining what has been working. What becomes clear from the orbital perspective is that a donor-based system that encourages wasteful spending and wasteful competition is a large part of—and potentially the primary cause of—the overall problems in development. A key part of the solution to this is to design financially self-sustaining projects. In the carbon credit model we used in water deployments, for instance, there is tremendous built-in financial incentive to keep things running effectively and to build routine maintenance, repairs, and replacements into the system, for the long-term viability of the project.

ORGANIZATIONAL STRUCTURE

Many humanitarian organizations have a hole in the middle of their organization chart, leading to a lack of guidance and mentorship at the lower levels of the organization. The lower levels, meanwhile, are typically staffed by enthusiastic and talented but poorly paid, inexperienced young people.

Development work is usually hard, messy, uncomfortable, and frustrating, and it typically pays less-than-competitive wages. Many people join humanitarian organizations because they want to have an impact, get some experience, and then move on to more mainstream jobs. This is understandable, but it makes the development field somewhat flat. It also tends to create organizations without a sufficient number of midlevel people who can serve as mentors.

Moreover, individuals who choose to stick with a career path within these organizations tend to change location every few years, because many organizations want their people to innovate, implement, and move on to do it again. These organizational stars may arrive on a new project in a new country and control a lot of money, but they may not have a lot of knowledge about or experience with that country. This can lead to inconsistency in approach, process, and performance and a loss of corporate knowledge. There is no real system in place to reward people in their careers if, for instance, they spend a significant amount of time in a Rwandan village learning about a real Rwandan family and its real needs.

A related issue is that a lot of people in funding organizations, talented though they may be, may not be highly experienced in the area where they are now expected to spend money. And when this lack of specific subject-matter expertise is coupled with a lack of rigor in the feedback loop of development projects, the people controlling the purse strings may not be able to discern noise from fact.

BUREAUCRATIC INERTIA

If most people are in development work to make an impact, yet the impact is lacking, as we have discussed, why do these issues still exist? We have seen the void in mentorship and the organizational structure that leaves a middle management gap. Corporate memory is often fleeting, and this may also be true of those who don't stick around long enough to discern the issues. But in organizations with good corporate memories, my guess is that a lot of those experienced people do realize that the system is broken and would like to change it but are faced with a kind of bureaucratic inertia. It can be very difficult to effect real change in any large organization, and very easy to get bogged down in the bureaucracy. Even those at a fairly high level within an organization will likely find it difficult to make major policy or procedural changes. A common frustration in large bureaucracies is that the people who know that the system is broken and know what needs to be done to fix it do not have the influence necessary to implement change.

The fact is, most large bureaucratic institutions simply do not take an orbital perspective. People at the top try to maintain a big picture perspective on the things within their control, while people at the lower levels concentrate on the worm's eye details of their specific projects. The people at the top must work within the confines of their present budget cycle, which does not encourage the long-term focus required for the orbital perspective. Likewise, the worm's eye view at the lower level is affected by the perverse incentives we've discussed. Most importantly, the synthesis of these two perspectives, which is the hallmark of the orbital perspective, rarely occurs.

For instance, the latest big picture thinking puts a very high value on community participation. The argument is that effective development occurs when developing communities take ownership of and participate in their own development. This makes perfect sense, and is probably often the correct strategy. But forced community participation across the board as the norm is shortsighted and ineffective, and if we zoom in to the worm's eye details, we see that some things about it don't make sense. For instance, it may cost between $5,000 and $30,000 or more to survey, drill, and install a hand water pump in a sub-Saharan village.[3] A community participatory development model would have the village form a water committee, raise some money from the local community to pay for future maintenance, and train one or more people in the village to be responsible for maintenance of the pump.

But this model doesn't work very well in impoverished communities. First, raising money from the destitute poor in the name of community involvement creates an impossible burden on them. In addition, this model requires people who are the least prepared to deal with the inevitable problems that will arise to take full responsibility for systems they are at best superficially trained in. It is estimated that from 30 percent to 80 percent of community-managed pumps are broken in the first couple years of operation, and most of them never get repaired.[4] This can set the system up for failure, and it really isn't fair. Most people in developed countries are not responsible for maintaining their water utilities; these systems are installed by and maintained by professionals. Expecting something different from impoverished people in developing countries represents a failure to synthesize the big picture perspective with the worm's eye view, which can lead to failure on the ground, with the result that nothing really changes.

As we've seen, however, this is not necessarily the fault of the well-meaning people who are attempting to solve a local problem. Instead, it is a result of bureaucratic inertia, an inability to change the way things are done. The incentives surrounding formal and informal institutions are, for the most part, not coordinated with the impacts they are seeking, and the way people are rewarded within their careers and their organizations is not necessarily aligned with the best way to overcome the challenges their organization is seeking to address. Such incentives are rarely aligned in an optimal way to create big picture, long-term impact, yet bureaucratic inertia makes it very difficult to change these things.

DESTRUCTIVE COMPETITION

One of the things that really surprised me when I started in development work was the level of competition even among organizations that have similar goals and want to solve similar problems. I naively assumed that all humanitarian organizations would see the overall cause as more important than individual organizational goals or the individual career goals of the people within them. Instead, it has become apparent that there are many factors that limit the good being done by organizations, and one of these factors is destructive competition.

Humanitarian organizations all compete for the same grants, donors, and attention, and there is a finite amount of money available. Generally, competition is a good thing, separating ineffective and substandard organizations from organizations that are really making a difference. Healthy, honest competition can lead to an improvement in the ability to develop and implement solutions. The problem comes when an organization makes itself look better while dishonestly discrediting competing organizations' accomplishments.

Nongovernment organizations are not filled with cutthroat businessmen, but they are incentivized to talk up their successes. To get donor funds, they have to highlight what they do best and play down, if not totally ignore, their weaknesses. At some level, too, they can end up disparaging other organizations—as too small, too inexperienced, even too underfunded to succeed. They may even downplay or undermine the successes of others, regardless of whether those successes actually move the planet toward the solution they are working toward themselves!

Rational, fact-based critique is a critical aspect of effective development. But secretive, gossipy criticism or undisclosed bias is something different and can be very destructive. We're in some sense used to this in the business world, where competing brands prop up their own image at the expense of competitors. But in business, consumers will quickly discount advertising if a product turns out to be substandard. In the humanitarian sector, on the other hand, because of limited rigor and lack of effective feedback mechanisms, it is sometimes very difficult to discern fact from fiction, information from noise, and this type of behavior can last longer and do considerably more damage. Unfortunately, too many factors prevent the distinguishing of the good guys from the bad, and therefore this kind of competition does not lead to improved quality but rather to a waste of resources.

Though it was not true during the Chilean mine rescue, destructive competition can exist even during disaster response. Some organizations that show up to a disaster area can be very territorial. They may have a particular view of the right way to do things, which aligns with their niche capabilities. Sometimes this focus on what they know, what they're comfortable with, and their previous experience masks the bigger picture and obstructs the competing approaches of other organizations. Even in that relief environment, there are organizations that are competing with each other to demonstrate that they've done the job better than the rest.

NATIONAL INTEREST

Destructive competition can also affect how nations interact within the global community. Looking at our planet from the orbital perspective reveals a collection of contiguous landmasses separated by water. But as we know, these landmasses are made up of a series of sovereign nations, and these nations all tend to act in their own individual best interest—or at least what the national leadership believes is the nation's best national interest. Cooperation occurs when the national interest of two or more nations align. I believe that most nations are open to acting in a way that furthers the common good of the planet, but when nations are faced with a decision between acting in the best interest of the world or the best interest of the nation, national interest will win out every time. This is one of the challenges to be faced in doing something inter-national.

The International Space Station provides enormous benefit to the entire world, and the fifteen nations of the ISS partnership rightfully take pride in their contribution to the greater good of the global community. But what enabled each of these nations to join the effort in the first place was a conscious decision that the effort was in their best national interest.

Likewise, numerous decisions every day guide a nation's path and progress. Some of these decisions put a nation on a good path and some do not. Some decisions are made by individuals who have acted honestly, responsibly, and with integrity, and some are made by individuals with hidden agendas, seeking personal gain or political power. National decisions are influenced by myriad factors, including politics, personal agendas, pride, ego, altruism, and compassion, but I believe that the decisions that nations make are affected by the same factors that affect individual decision making, with one apparent key difference: individuals are considerably more likely to act selflessly.

When an individual is faced with a decision between what is good for him or her personally and what is for the greater good, it is conceivable that a person could sacrifice for the greater good. This happens all the time. Nations, on the other hand, seldom if ever act exclusively in the best interest of the global society, if it comes at the expense of their own national interests. After all, politicians are generally expected to further the national inter-est. In cases where a nation acts in a manner that benefits other countries or the world, to its own apparent detriment, I would be willing to bet that this is really a sacrifice of *immediate* best inter-est in favor of that nation's long-term, big picture best interest. In other words, if we pull back and consider the implications from a global scale and over a long time frame—which, by this point in the book, we should all recognize as the orbital perspective—we can see that the chosen course actually was in the best interest of the nation.

If we remove the undesirable personal motives of individual politicians and assume that decision makers are acting respon-sibly, honestly, and with integrity, then a decision about whether to support something that furthers the greater good of the global society should ask the following questions:

1. Is the decision in our best national interest?
2. If we project the predicted outcome out further in time, is it now in our best national interest?
3. Are we as a nation willing to sacrifice our national interest for the greater good?

If the answer to question 1 is yes, then the nation will likely decide to follow that course. If the answer is no, then the second question should be considered. An answer of yes to question 2 should again prompt the nation to follow that course, but an answer of no should lead to question 3, and I believe that nations that get to question 3 will say no nearly every time.

What this means is the decision makers either didn't pull back far enough to see the benefit when considering question 2, or there really is no benefit and the nation is not equipped to make that type of sacrifice. On the other hand, I believe individuals are equipped to make that type of sacrifice, and this is why I believe that a collection of individuals is more likely than a collection of nations to make real progress toward solving our biggest challenges.

In the early U.S.–Russia space collaboration and the ISS program, it was the personal trust among individuals that carried the programs through a doctrine of exclusively national interest. The tendency of nations to act in their own best national interest is a barrier to international collaboration, but in some cases this can be overcome through the trust that develops within an international program, which is founded on personal relationships.

Patriotism and loving one's home country is a wonderful aspect of humanity, but love of country needs to occur in the context of the rest of the world—that is, with an orbital perspective. Truly loving something means seeing it clearly and loving it in spite of its faults and imperfections. No country is perfect, and when national interests don't coincide and countries take the pursuit of national interest to the extreme, war breaks out. War is humanity's greatest failure.

However, if we remove the stressors caused by the requirement to pursue national interest and instead interact on a person-to-person, organization-to-organization level, then a cause itself can gain greater influence, enabling remarkable collaboration that does not require the establishment of deep personal relationships.

In the next chapter we'll look at several movements and initiatives that are synthesizing the orbital perspective and the worm's eye view, making vast amounts of information accessible and bringing large groups of people together to work toward common causes. In some cases, collaboration is not facilitated by long-term personal relationships. Instead, the bond is the common cause, the shared experience, and in some cases simply being given the permission and the means to help. In all cases, however, groups of individuals working together toward a goal have been accomplishing far more than any individual—and perhaps any organization—can alone.

CHAPTER NINE

Mass Collaboration

✦ Our global society is producing data at an exponentially increasing rate. Data enables decision makers to determine the most effective ways to address their most critical challenges. Countries, cities, and communities that can identify key issues, determine where their most vulnerable citizens are located, and understand the needs they face are equipped to accurately determine how to best apply limited resources to achieve solutions. But data by itself is insufficient. We also need tools to analyze that data and to translate the analysis into more effective and targeted approaches that can dramatically improve society's ability to meet our grand challenges.

Fortunately, along with the dramatic increase in our ability to produce data, there also have been recent developments in the power and ability of tools to analyze, make use of, and communicate the insights of that data worldwide. Among these developments is the use of crowdsourcing to process data in what is commonly known as a hackathon or codeathon. These mass collaborations are organized around various themes and were born out of a public–private initiative called Random Hacks of Kindness. Other ingenious mass collaboration tools include ReCAPTCHA and Duolingo, which tap the previously unutilized, distributed efforts of millions of people to perform massive tasks—often without users knowing they are taking part in a mass collaboration. These efforts share in common a desire to make use of massive data stores to create social and/or environmental good, and such data sets may come from anywhere, including NASA.

TO SERVE

Conservationist Dan Irwin was fresh out of grad school in 1993 when he was tasked with slogging through the dense jungle of northern Guatemala. His job was to map the newly created 2.1 million–hectare (8,800-square-mile) Maya Biosphere Reserve, survey its communities and resources, and develop accurate data that would inform critical decision makers about the rain forest. He was collecting data by hand, using the equipment of the time: a notebook and a large GPS receiver reminiscent of the brick-size mobile phones of the 1980s. In all, Irwin spent more than a year walking ten to fifteen miles a day through dense jungle, encountering hazard and risk at every turn. There were mosquitoes carrying malaria, severe blisters that required Irwin to be evacuated by mule, and the deadly barba amarilla (yellow beard) snake, which claimed the life of one of Irwin's friends working nearby.

Each month, Irwin spent twenty-two consecutive days in the field, followed by eight days in town. During one of his final eight-day cycles in town, Irwin met Thomas Sever, NASA's only archaeologist and a pioneer in remote sensing technology. While Irwin was mapping the surface of the jungle, Sever was using infrared satellite imagery to detect ancient Mayan ruins buried beneath it. As they were talking, Sever pulled out a map and rolled it out on the table.

"I about fell off my chair," said Irwin. He was looking at a view of the reserve from space, created through satellite imagery. "I just spent the whole year walking through the jungle, and here I could see just about everything I had mapped, in such unbelievable detail." Suddenly, Irwin's worm's eye view from the ground shifted to the orbital perspective from space. The imagery provided important data of the type Irwin had spent an arduous year collecting, but Irwin saw equal value in its ability to communicate the story to those most directly affected by the environmental issues he was studying.

Soon after his meeting with Sever, Irwin packed a slide projector on a mule and began traveling from village to village, showing the images to the people who lived there and presenting this broader, deeper perspective to villagers and national decision makers alike, expanding their field of vision. "From the ground, the forest appeared to the villagers to go on forever. The satellite images showed them the truth."

The truth was that the agricultural frontier was moving toward them from the south, and the Mexican border, which was heavily deforested, was not too far to the west. Irwin was able to show the local people what was really going on in their vicinity. "It catalyzed their opinion and their spirit of the purpose of this reserve and the importance of the rainforest."

Irwin's experience was a concrete example of the orbital perspective: if you step far enough back from any problem it becomes everyone's problem, and everyone gets in the picture. The satellite data and imagery was doing more than connecting space to Earth. It connected space to village, providing incredible context and the ability to prioritize and manage efforts.

SERVIR: SCIENCE IN ACTION

The experience in Guatemala broadened Irwin's understanding, but it also changed his life. In 1999, Sever invited him to work at NASA's Marshall Space Flight Center in Huntsville, Alabama, where Wernher von Braun had led the design and development of the Saturn V rocket that launched humans to the moon. Meeting the scientists and researchers at Marshall, Irwin was struck by the innovative and important work everyone was doing, using satellite data for such things as thunderstorm prediction and hydrologic modeling. But Irwin sensed a missing link between what his colleagues were doing and what would be done with their work. "Nine times out of ten," he said, "their end goal was to publish their research and present it at a conference."

Irwin was not satisfied. He knew from his experience in Guatemala that there was a bigger need, and a big demand to apply this cutting-edge scientific research to real world problems. People around the world could benefit from access to this information, and Irwin was determined to bridge the gap.

Ten years after taking the first steps through the Maya reserve, he got his chance. He was invited to meet with the environmental ministers of the seven Central American nations, who were eager to collaborate across political borders to tackle environmental challenges that had no regard for these boundaries. The ministers knew of Irwin's work mapping the lands and resources of the Maya reserve, and they recognized the opportunity to incorporate NASA's data and apply it regionally.

Central America represents only 0.1 percent of Earth's landmass, but the region experiences natural disasters and environmental pressures disproportionate to its size. Volcanic eruptions and earthquakes are not uncommon. The region is challenged by algae blooms in the Pacific to the west and the Caribbean to the east, deforestation from slash-and-burn agriculture, and air pollution. Faced with these realities, the environmental ministers recognized that, although they are seven countries, they are really one region, and they needed a practical way to cement this perspective. Looking from the point of view of space was one way of accomplishing that.

Out of this discussion came Irwin's idea for *Sistema Regional de Visualización y Monitoreo* (Regional Visualization and Monitoring System), or SERVIR. It seemed to Irwin to be the perfect name for the program. "It's the Spanish verb 'to serve' and has so many meanings on a personal level for humanitarian work—the civil servant has the notion of serving, the NASA spirit has that notion of serving. It all came together under the banner of serving."

SERVIR integrates satellite observations, ground-based data, and forecast models to monitor and forecast environmental changes, enabling scientists and decision makers to better address a range of issues, from climate change and disaster management to agricultural development, biodiversity, and conservation. NASA would provide and coordinate space-derived information, data, and analysis, and the U.S. Agency for International Development (USAID) would be brought in to provide regional knowledge and infrastructure, and to support a regionally focused local organization with people on the ground—a truly collaborative, sustainable, and scalable model. SERVIR brought together two organizations chartered to improve life on Earth—a space agency and a foreign assistance and development agency that works with more than one hundred countries—to provide the orbital perspective to those with the local, worm's eye view on the ground.

A SERVIR coordination office and rapid prototyping facility was soon set up at Marshall Space Flight Center, and regional SERVIR hubs were established through existing organizations around the world, connecting space data to the needs of the development and crisis response communities. The first hub was established at the Water Center for the Humid Tropics of Latin America and the Caribbean, in Panama. People were hired from each of the Central American countries to ensure a regional presence and to provide each of these nations a seat at the table in dealing with their issues.

SERVIR next expanded to serve twenty-one African countries, with a hub at the Regional Center for Mapping of Resources for Development, in Nairobi, at which point Irwin knew that the project had the ability to spread worldwide. A hub at the International Centre for Integrated Mountain Development, located in Kathmandu, quickly followed to serve Bangladesh, Bhutan, Burma/ Myanmar, and Nepal. There presently are additional plans to expand to Southeast Asia and West Africa.

Remarkably, NASA and USAID did not unilaterally establish these hubs, though they provided the push. NASA provides valuable space-based data, which the hubs process and disseminate. Consistent with the primary SERVIR goals of training, capacity building, and scientific uptake, NASA and USAID provide support to the hubs but the people in the region take ownership. In addition, the hubs are organically replicating themselves. The Central American hub helped build the East African hub. The Africans consulted with the folks in Nepal to show them what they were doing, and surely they will help the next region to come online. Each region is sharing knowledge to build a global network. For the most part, this is accomplished through commonly available social and communication platforms such as Twitter, Facebook, or Skype.

For Irwin, the most rewarding result is that SERVIR catalyzed the involvement of so many elite scientists and researchers around the world. "It's been a really nice example of scientific diplomacy," he said. "It's been a sharing and a two-way street of knowledge exchange. We may have some of the satellite knowledge and the applications knowledge, but it's the people in the field that understand how their system works, and we're bringing those two together."

Indeed, one of SERVIR's fundamental principles is to listen to and understand the unique needs in each region. The SERVIR hub model was very valuable in translating NASA's data to government ministries and ultimately to an end user on the ground, such as a farmer in the field. This trickle-down approach was the most cost-effective way to get deeper and deeper and to provide more and more valuable, actionable information to decision makers at all levels—from space to village. In some cases, the big picture perspective provided by this data has prompted governments to transform their resource management. Regional disaster relief coordinators can prioritize emergency response, and environmental decision makers can focus their stewardship to better protect resources.

Mass collaboration on utilizing space-derived data also has improved the ability of people in different regions to understand each other. The ability to look at Earth together helps us to see our world with fewer boundaries, and really catalyzes a dialogue that leads to working together. "We're seeing more and more that the fires in the Yucatán Peninsula affect the air quality here in Alabama, or the atmospheric brown cloud in Asia is affecting the glaciers, and the glaciers are melting faster," Irwin said. "There are a lot of problems out there, but I'm also hopeful . . . because the conversation is occurring. And people have access to information and there's transparency in what's going on."

When we pull back in both the literal and figurative sense to the perspective provided by satellite imagery, things become clear. Not only do we see the pieces of the puzzle and who has them, but also we see where the pieces fit. SERVIR is one effort to put these pieces together, leading to many improvements to life in the regions served—everything from flood modeling that provides earlier warning and saves lives in low-lying Bangladesh, to a system that detects fires in remote regions of Nepal and Bhutan, before they get out of hand, to dozens of additional capabilities. SERVIR stands as a shining example of the power of a willingness to work together to put the puzzle together, recognizing that Earth is a small town with many neighborhoods in a very big universe.

SPACE APPS CHALLENGE

In creating SERVIR, Irwin synthesized the big picture with the small picture and turned both into action, taking the big picture orbital perspective of space-based information and applying it to the worm's eye detail of work being done on the ground. That same spirit of cooperation is being used to change the way we innovate.

The industrial age model of innovation was that organizations, governments, or businesses would pull together a large team, create a business plan or program road map, establish a budget, and then systematically roll out the innovation in logical, incremental stages. This approach worked well for decades, and still does—it got us to the moon and back safely. This type of process was and still is very appropriate for programs where failure is not an option, where the smallest mistake can lead to catastrophe.

Nonetheless, these tried-and-true innovation processes are increasingly being supplemented by the developing processes of the information age. More and more innovation is thriving in a bottom-up, decentralized structure, and today's innovation thrives in environments that accept failure as an integral part of the process. Meaningful global engagement and collaborations now routinely happen across industries, disciplines, borders, cultures, and regions. Mass collaborations take on various themes and flavors, but they generally are defined as the engagement of a large group of diverse participants to address specific challenges collectively and collaboratively.

For instance, TechShop, which began in Silicon Valley and now has multiple locations throughout the United States, is essentially an industrial machine shop that people can use on a membership basis. The concept is to democratize manufacturing. Startups and social entrepreneurs who want to create rapid prototypes of innovative ideas, or who already have a proven product but lack the capital to build full manufacturing facilities, can use these types of membership-based facilities to get their businesses off the ground. This model can have a tremendous positive impact on entrepreneurs, providing the manufacturing base to get products developed and produced without a huge up-front investment.

In April 2012, I helped facilitate the first-ever International Space Apps Challenge at a TechShop in San Francisco, which was a host location along with twenty-five other venues in seventeen countries on all seven continents, from Canada to Kenya to McMurdo Station, Antarctica. Space Apps was the brainchild of NASA's Open Innovation team, with the support of collaboration agency SecondMuse and digital humanitarian organization Geeks Without Bounds, and focused on using space data to improve life on Earth and in space.

This kind of undertaking is not new for NASA. In fact, NASA's response to the Apollo 13 mission, in which the spacecraft was crippled by an onboard explosion, might be considered its first hackathon and a first, real-life space apps challenge. During that historic mission, with the lives of the three crewmembers hanging in the balance, NASA had to solve a massive challenge with limited resources and limited time—a solution had to be hacked. As NASA Open Innovation's Ali Llewellyn put it, "An unexpected group of people came together in an unexpected way to solve an unexpected challenge. Our goal was to keep that spirit alive in the International Space Apps Challenge."

NASA's goal with Space Apps was to tap into and apply collective genius to solving problems in our world and in space exploration, to engage thousands of people around the world in NASA's mission. Using publicly available scientific data, participants would look at issues such as weather impact on the global economy or the depletion of ocean resources. However, in contrast to the traditional model of innovation, which works to avoid failure at all costs, Space Apps was founded on the principle that more experimentation means more failure—but also more opportunity to learn.

Like SERVIR, Space Apps stemmed from the desire to explore the possibilities offered by the terabytes of NASA-produced data, which NASA wanted to make available to the public, to universities, and to citizen scientists around the world. But as Dan Irwin had discovered, there's a big difference between providing data and presenting data in a user-friendly way that the public can actually use for good. The Space Apps event—at the time, the largest mass collaboration ever attempted—brought together more than two thousand registered participants focused on seventy-one challenges related to data management and presentation. The challenges were grouped into four broad categories: open source software, open hardware, citizen science platforms, and data visualization. The forty-eight-hour event resulted in more than one hundred solutions, all of which were developed in a completely open source environment. Volunteers produced amazing data visualizations, software applications, open hardware solutions, and great educational outreach products. The event was a huge success, and really illustrated the power of taking a big picture orbital perspective.

This inaugural challenge marked NASA's first large-scale mass collaboration for the specific purpose of promoting innovation through international collaboration, and it has been repeated each year since 2012, growing in size, scope, and results. The 2014 challenge, for instance, involved 318 collaborating organizations, in ninety-five cities and forty-six countries, again on all seven continents. It brought together 8,196 registered participants to address sixty-seven challenges.

THE PINEAPPLE PROJECT

Samantha Snabes, who at the time was the deputy strategist for space life science at the Johnson Space Center, had been inspired to join the JSC chapter of Engineers Without Borders after hearing the details of the collaboration that brought the Chilean miners safely through 2,300 of hard rock to the surface. "That's where the wheels started turning for me about the connection between space and society," she said. "It started giving me the words and example of how the orbital perspective translates to Earth."

After joining the organization, Snabes had found herself on multiple development trips to Africa and Central and South America. As part of this work, in 2011, Snabes visited L'Esperance orphanage, where I worked as part of a Manna Energy Foundation–Engineers Without Borders team. Nestled in rolling hills overlooking the shores of Lake Kivu in rural western Rwanda, L'Esperance was run by a Guatemalan named Victor Monroy, an amazing person who proves that you don't have to go into orbit to have the orbital perspective. Monroy's vision for the orphanage was to build a self-sustaining community that could meet the orphans' basic needs while also providing employment and funding higher education for the children as they became adults. To that end, Monroy planted a series of tropical fruit orchards. The orchards would help feed the children and workers at the orphanage, and excess fruit could be processed through on-site drying facilities, creating an export business to raise desperately need revenue.

Inspired by what she saw at the orphanage, Snabes resigned from her official duties at NASA and committed herself to helping the orphanage achieve self-sufficiency, supporting efforts to design and build fruit driers that would work at the elevation and climate conditions of the orphanage. This effort required bridging the disconnect between technology and the needs of the poor. At the first Space Apps Challenge in 2012, therefore, Snabes took the lead on what became a global effort to develop a way for farmers to determine the best crop to plant, given local conditions. The challenge was called the Pineapple Project.

The goals of the Pineapple Project were applicable not just to the orphanage but also to many in developing countries who could improve their diet and livelihoods by cultivating fruits. The success of such efforts depends on knowing the fruit type and variety that will grow in a particular region, and the challenge of the Pineapple Project was to gather existing agricultural, climate, and topography data, design a matching algorithm to choose the right type of crop to plant based on these parameters, and make this information widely available via a website or a mobile app. The goal was to create a tool that would tell farmers not just which variety of fruit to plant but also how much to plant, decreasing the likelihood that farmers would spend all of their available resources planting the wrong crop—only to find, after a long cultivation period, that it didn't work. In many cases, the people making these critical crop decisions were subsistence farmers, and failure to plant the correct crop could lead to an inability to feed their families. Challenge participants were supplied with temperature, soil, light, elevation, and precipitation needs for common varieties of guava, mango, papaya, and pineapple, and were encouraged to identify additional data sets or resources that might be useful in the effort.

The Pineapple Project held great agricultural promise, but it also served as a great case study of the interaction and dynamics that take place during such a collaboration. At Space Apps San Francisco, I saw teams from around the world come together, assign tasks, engage in periodic videoconferences, and develop the start of a very useful tool. Teams worked around the clock, handing off tasks around the globe. The team in Nairobi handed off what it was working on to the Santo Domingo team and then tagged back up later, after they got a little rest. According to Snabes, "The thing that was most interesting about the Pineapple Project was, when we were all finished . . . we realized we had gone the entire hackathon, hacking around the world, and nobody knew anyone's official occupation. They just knew what they were working on for the Pineapple Project."

This is one of the crucial components that enables a hackathon to be so innovative and productive. Participant roles are not assigned based on title, stature, or even experience. Instead, people work on what they are passionate about, and responsibility is assigned or assumed based on individual desire and motivation.

Although a fully functioning prototype was not developed by the end of that first challenge, the teams did produce a vastly improved scoping of the problem, identifying where the holes were and what needed to done. Even without a finished product, however, the progress on the Pineapple Project was so impressive that the combined team from around the world was awarded the Global Award for "Most Disruptive." The project continued on through subsequent hackathons and was eventually invited to participate in an accelerator program. Today, the project resides under a nonprofit organization, which allows the project to be continually worked on and coordinates the efforts of volunteers all around the world.

ISERV

Dan Irwin also was involved with the International Space Apps Challenge. From the beginning, the heart of the SERVIR mission has been collaboration. Therefore, his team immediately recognized the value and opportunity for direct civic engagement and collaboration at the heart of the challenge, and Irwin led the charge to facilitate Space Apps events in the African, Himalayan, and Mesoamerican regions served by SERVIR.

One of the challenges at Space Apps Kathmandu 2014 was to support a SERVIR spinoff project known as the ISS SERVIR Environmental Research and Visualization System, or ISERV. ISERV is a camera that acquires Earth images through a small telescope attached to a window of the International Space Station's Destiny laboratory. Coincidently, I installed some of the hardware that interfaces with the camera, during my 2011 mission.

Since its installation, ISERV has received commands from Earth to acquire many tens of thousands of images of areas threatened by floods, landslides, forest fires, earthquakes, and other natural disasters. The challenge for the Space Apps collaboration in 2014 was to provide easy access to and visualization of this georeferenced imagery data. Three robust mobile apps were developed during the challenge, enabling people to visualize and map the data on a smartphone or tablet.

The ISERV project is one of many helping to fulfill one of Irwin's goals in starting SERVIR: to help developing countries deal with mini-disasters, such as a landslide in Uganda, a flood in Namibia, or a forest fire in Costa Rica. Such disasters may not receive the same media, satellite, or mass collaboration mapping attention as a large disaster, but they nonetheless have a huge impact in the communities, regions, and countries where they occur.

ISERV is designed to use the SERVIR network backward. Now, when a country has an issue, it can ask the ISS to collect needed data and send it down, making it available for the use of emergency agencies. In areas without fixed network infrastructure, or when infrastructure is threatened or destroyed, mobile apps help keep the data accessible. The beauty of this system is the two-way transfer of information. End users also are submitting imagery through the SERVIR network, and this is all feeding back through the app to decision makers.

RANDOM HACKS OF KINDNESS

The point of SERVIR and the Pineapple Project is to connect space-based data with worm's eye needs on the ground. Hackathons such as the Space Apps Challenge are the means of doing this, and are based on collaboration. In fact, this collaboration itself began as a plan by rivals to join forces to respond to global crises.

In 2009, representatives from three of the biggest tech companies at the time—Google, Microsoft, and Yahoo—began discussing the use of technology to improve crisis response.[1] At the time, there really was no substantial collaboration among these companies, who were fierce commercial rivals, but all agreed that some issues of global importance should take precedence over competitive business concerns. This was more than making sure people can have the Internet after a storm; increased connectivity would save lives. During natural or human-caused disasters, there's a need to communicate with the general public, to locate people in need, to coordinate first responders, and to provide information technology to coordinate all aspects of disaster response. Bringing together the respective strengths of each of these very capable organizations could really make a difference in reducing the hardship, suffering, and loss of life following disasters.

Soon, the World Bank was involved, as was NASA. The idea of creating a series of hackathons to bring together the formal and informal sectors was born, leading to a partnership between Google, Microsoft, Yahoo, NASA, and the World Bank and the first hackathon involving the U.S. government. In November 2009, the collaboration agency SecondMuse facilitated a crisis response hackathon called Random Hacks of Kindness, or RHoK (pronounced like *rock*), at the Hacker Dojo in Mountain View, California. Each of the tech giants brought its large developer community to bear and combined these with disaster response experts inside and outside of government and the massive data sets from NASA and the World Bank.

The event was a huge success. Developers came out in large numbers because the event provided an outlet to make a difference—many technologists found it frustrating to hear about a disaster, to know that they had a skill set that could provide valuable help, but to have no mechanism to provide assistance. RHoK was building a community of trust that could be called upon in times of crisis. It also built awareness about existing communications channels and provided a way to establish needed digital infrastructure, making it much easier to communicate and coordinate response when it is immediately needed.

RHoK, which started out as a crisis response initiative, grew in scope to include many other types of challenges. Today it is a self-sustaining global community of thousands of innovators in more than thirty countries seeking to develop practical, open source technology solutions to humanity's many challenges. RHoK's model begins by identifying and refining the problem definitions provided by local experts, ensuring that volunteer time is focused on solving real problems for real people.

GEEKS WITHOUT BOUNDS

A mass collaboration ecosystem and a whole series of private-public events has grown out of RHoK, including the International Space Apps Challenge. For instance, Geeks Without Bounds, founded by social scientist Willow Brugh and telephony expert Johnny Diggz, hosts an accelerator program and helps to facilitate Space Apps and other hackathons.

GWOB was initially founded to unite geeks for good and to connect maker and hacker spaces—"to basically become the help desk for the world," as Brugh put it. But the role of GWOB has evolved since its inception, and "now our aim is to bring good intentions to deployment." GWOB and organizations like it serve as the substrate that enables informal volunteerism to make a huge impact within formal government, corporate, and nongovernment structures. GWOB works to provide people with the information and resources they need to affect the world in the way they would like.

"It's frustrating that we have all these people who want to help with incredible skills, and the formal sector is closed off because they don't know how to make use of that capacity," said Brugh. "That is where Space Apps was, and continues to be, exceptional. NASA, as a formal organization, saw all the data they had, and their mandate, but that they were lacking [human capital] resources. So they asked for help, and got it, in spades. GWOB provided a bridge structure from the starting development at the Space Apps Challenge into full deployment, so the projects see the light of day."

Brugh also helped facilitate the initial Space Apps Challenge in San Francisco and was impressed with the collaboration and the results of the Pineapple Project, so GWOB invited the project into its accelerator program, which provides skilled mentorship and guidance on the development process. This is in keeping with the goal of the GWOB accelerator, which is to avoid wasting all the energy that goes into hackathons, to bring projects from a good idea and a good intention into actually being useful.

UNLEASHED CREATIVITY

When I first heard about hackathons, I couldn't understand why people would devote so much time, effort, and resources to what are usually other people's projects. What would make people willing to spend an entire weekend working for free? According to Brugh, however, participants don't look at it as work, but more like solving a puzzle. "People like to see how what they do can apply to things they care about and things that matter." In fact, when people *do* think it's work, the model often doesn't work. At the Space Apps Challenge, some of the visiting senior NASA managers, who were not directly involved with the design of the hackathon, thanked participants for working on challenges that NASA didn't have the budget to solve, demonstrating a misunderstanding of the event. Luckily, however, this didn't seem to discourage participation.

"It's productive when people have a chance to be creative, rather than you giving them a set task list," said Brugh. "You need to give them creative license or give them a way to learn something new. People want to be creative. You're there to collaboratively build something; you're there to try out something that you rarely get to do at work."

What is really interesting is that prior to this new approach of nonmonetized collaboration, there was a strong push within the federal government for monetized competitions. Competitions with monetary prizes certainly have proven to be an effective model for innovation, but Random Hacks of Kindness, the Space Apps Challenge, and similar efforts have introduced the idea that citizens want to participate not just for a prize check but also for other, even altruistic reasons.

In fact, one of the things that makes these government-led hackathons work is simply giving the public the permission to do something interesting and meaningful. According Nick Skytland, who led the Open Innovation team that conceived of the Space Apps Challenge, people are drawn in part by the opportunity to do something no one has ever allowed them to do before—to work with a government agency, to get involved in their local community, or to meet new people.

This is democracy in action, by the people and for the people. Democracy is meant to be participatory, and thanks to technology, it now can be participatory in a whole new way. Citizens can engage directly in their government at many levels. Although governments are focused on national interest, and the priorities of large bureaucratic organizations can be misaligned, they nonetheless can play a role in mass collaborations. In fact, due to shrinking budgets and the rising complexities of issues facing governments, it's probably more important than ever for governments to engage their employees and citizens in large-scale collaborative processes. Mass collaborations involving the citizens who are affected by government policy are a logical response to the pressure to do more with less and to do it with increasing transparency and openness. Ideally, citizens should have access to government services and information anywhere, anytime, and on any device. Unlocking data and making that data available to the public in a user-friendly way spurs innovation, improves quality of life and workforce productivity, and saves lives.

Although the results of a mass collaboration can make a significant impact, though, the most important results often come from the process of collaboration itself. Mass collaborations bring together groups that have never had the opportunity to work together before, revealing valuable connections and complementary skill sets. In addition to government agencies, businesses, nonprofit organizations, and citizens' groups all have made use of these powerful forums for innovation, and this trend is likely to accelerate as the reach, speed, and effectiveness of global connectivity increases and more and more people are added to the global conversation.

Mass collaborations provide an avenue for direct civic participation in government, direct stakeholder participation in humanitarian assistance and disaster response, and a direct connection between businesses and their customers. And once platforms for interactions are in place, it serves as a continuous conduit where a community can engage with, assist, and support those providing services, leading to an increase in efficiency and the possibility of rapid implementation of appropriate solutions. Community involvement is also likely to result in a much higher acceptance of and support for resulting solutions. Overall, open source mass collaborations provide a very cost-effective way to supplement the research and development needed to overcome our biggest challenges.

RECAPTCHA AND DUOLINGO

The power of mass collaboration lies in its ability to amplify and aggregate relatively small investments of time into something large and meaningful, but hackathons are just one example of this. Mass collaborations are starting to happen all around us, sometimes without our awareness. Take, for example, ReCAPTCHA.

Most of us are aware of CAPTCHAs, even if we don't know what they are called. The Completely Automated Public Turing Test to Tell Computers and Humans Apart, designed by researcher Luis von Ahn and others at Carnegie Mellon University, is that distorted, slanted, and otherwise modified set of letters and numbers you sometimes have to type before submitting online forms. CAPTCHAs are designed to prove that you're a human, because computers are not yet able to decipher those squiggles, preventing such things as ticket scalpers writing programs to automatically buy thousands of tickets that they will then resell illegally.

ReCAPTCHA is an offshoot of this project, stemming from the realization that humans type about two hundred million CAPTCHAs into Internet pages every day—totaling more than 500,000 hours, if typing a single CAPTCHA takes ten seconds. But in typing that CAPTCHA, the human brain is doing something a machine can't. And that human capability is now being used to help digitize books.

Digitizing old books usually involves scanning the books and then converting the images into text using optical character recognition algorithms. Unfortunately, computers can't recognize all of the words, and generally the older the book is, the harder it is for computers to decipher the words. ReCAPTCHA takes the images computers can't recognize and uses those words as CAPTCHAs, using humans to identify the words the computer couldn't. If this unknown word is presented to a number of humans, and all identify it as the same word, the word is now accurately identified. As of April 2011, this process has digitized about one hundred million previously unrecognized words every day—about 2.5 million books a year—all through the efforts of 750 million distinct people typing CAPTCHAs on the Internet.[2]

The same team took a similar approach in designing Duolingo, a foreign language teaching program that, when used, simultaneously translates portions of the Internet. The system uses real content and is as effective at teaching a new language as leading commercially available language software programs. But while it is teaching, it also combines the translations of many beginning language students to get the results that would be produced by a single professional translator and that, surprisingly, are just as good. Duolingo ingeniously provides two simultaneous benefits: it provides free language training for the poor, who have the greatest need for and reap the biggest benefit from it, and it is translating the Internet into multiple languages, again with tremendous benefits for the poor.

Ahn noted that some of the biggest human projects in history, including putting a man on the moon, all employed roughly one hundred thousand people, because before the advent of the Internet, it was not feasible to coordinate the actions of a greater number. "If we can put a man on the moon with one hundred thousand people," he said, "imagine what we can do with one hundred million people."

ReCAPTCHA and Duolingo are examples of collaboration on this kind of massive scale, ingeniously making use of human effort that otherwise would be wasted. These initiatives and many more like them, which have recently been made possible through improved information technology, will also be critical to helping us overcome the world's biggest challenges. Moreover, if we tie all these developments in mass collaboration with better accountability, through pay-for-performance models and the improved data feedback made possible by the widespread use of inexpensive sensors, we can see vast improvements in the effectiveness of development work worldwide. In the next and final chapter, we will look at what significance and opportunities these collaborative capabilities present for the trajectory of our global community.

A Web of Trust

Like the U.S.–Russian space program that led up to it, the planning and construction of the International Space Station required the partners involved to overcome some unique challenges. The collaboration brought together fifteen nations with different bureaucratic and political processes and differing national objectives and interests, geographically separated on three continents. Many key personnel didn't share a common language or culture, and there were complexities related to intellectual property issues. Because almost all space technology has a potential dual military use, the partners also had to work within the overly restrictive mandates of the United States' International Traffic in Arms Regulations (ITAR).

In addition to these cultural, national, and geographic obstacles, the partners had to collaboratively design a massive, complex space station and figure out how to run it. They had to agree on who would supply which parts and capabilities to the station, how the day-to-day operations would be run, both in space and in interconnected mission control rooms all around the world, and they had to collaborate on the scientific research that would be conducted on board. Finally, they had to collaboratively execute all those plans and designs. For the most part, this collaboration was enabled through international treaties and formal contracts, which required documentation of the smallest details, and the nature of the program itself dictated that the processes and procedures established should be designed to last for a long time.

Despite the obstacles, however, the collaboration that emerged from this effort built arguably the most complex structure in history, and the partnership itself was so effective and had such impact that it was nominated for the Nobel Peace Prize. But the collaboration would never have worked without the trust that arose in personal relationships among team members. Building this trust required a long process of small, incremental steps, which gradually built in complexity and scope. In some cases, initial deep mistrust and competition had to be overcome, particularly because the United States and the Soviet Union had once been on opposite sides of the Cold War and competitors in the space race. But among all the partners, the "secret sauce" that made the partnership work was trust. Without exception, everyone I spoke to who had been involved in the partnership said that direct face-to-face relationship building created the foundation of trust that built the space station.

There are many similarities between the collaboration that was required to build the ISS and the elements required to collaboratively solve the problems facing our world. As with the space station project, collaboration on solving our big challenges will require a movement toward aligning discrete efforts. We need to develop mechanisms to engage individuals and organizations seeking to do good into a unified, coordinated effort. Part of this process is to increase the amount of data we share, so that we can pull back to the orbital perspective and develop a complete picture of the playing field. Everyone involved with tackling a global challenge should know who is doing what and where, where there are unmet needs, what's truly working, and what's not. Everyone should be engaged in dialogue about appropriate solutions.

In addition to sharing more data, we also need to bridge the gap between the data that is available and the data that is readily usable, in order to put actionable information in the hands of decision makers at all levels. Emphasis should be placed on developing innovations to utilize underutilized human capital by lowering the barriers to help. In short, we need to think big and long—we need to develop an orbital perspective. This will require reducing our focus on quick fixes and Band-Aid solutions and tearing down barriers to taking an orbital perspective approach, such as short budget cycles, perverse incentives toward the new and shiny, and unsustainable funding sources. And we need to circumvent or make obsolete those barriers we are unlikely to overcome in the near future, such as corruption or national objectives that don't coincide with one another.

PROJECT X

There are, however, some important differences between the effort to build the space station and the components that will be required to manage global problems. To highlight some of those differences, let's imagine an international program with the goal of providing clean water to every human on the planet by 2030. We'll call this imaginary program the International Safe Water (ISW) program and model it after the ISS program. To create such a program, we would bring together a dozen or so nations, sign treaties, form working groups among the partner nations, integrate major water-related nongovernment organizations, and collaboratively design a comprehensive plan. This type of collaboration would require the relationship building and trust building that occurred in the ISS program, but of course would incorporate today's information technologies.

But we could take another approach to the same effort, which we will call Project X. Instead of a formal international coalition, this would be a less formal, loosely managed, community-driven effort. For this project, we would provide incentives to capture as much data related to the issue as possible, provide incentives to convert that data into something usable and actionable, and where appropriate, encourage organizations to partner and share resources on both long-term and short-term projects.

Both the ISW and Project X programs would involve a big important goal, with lives at stake. Both have a potential requirement to collaborate across long distances and a need to overcome language and cultural barriers. Both need to establish a level of trust and build something that will provide long-term benefits.

Project X, however, would not necessarily require formal contract mechanisms across the board or need to operate within a formal political system. There also might be no requirement to carefully measure individual contributions to ensure that formal agreements are fulfilled. There wouldn't necessarily be a need to ensure that everyone involved is financially or otherwise compensated for all work. There potentially would be fewer requirements to work within a proprietary intellectual property framework or to follow a hierarchical command-and-control structure. Although both programs would aim for long-term benefit, only the ISW program, following the ISS model, would require that collaborations stay in place for a long time to ensure that happens. On Project X, typical collaborations could come together for a short time and stay together only for the duration of the task at hand. As a result of all these differences, Project X would have a potential for explosive exponential growth that is not found within formal bureaucratic structures.

PROVISIONAL TRUST

Like the ISS program, the ISW program would require long-term, deep personal trust. Project X, on the other hand, would probably have significantly fewer requirements to build personal, trusting relationships, and when those relationships were required, members of the Project X community would be able to make use of interactive technology that didn't exist when the ISS program was developing. Most of the time, a more moderate level of trust would be sufficient to connect individual pieces of the larger puzzle.

The ISS collaboration was built on the deep personal trust that comes from having meaningful relationships with other team members, but Project X would be based on provisional trust, which emerges only for as long as it's needed. In both cases, trust leads to the belief that each side will do what they say, provide what they promise, and not take advantage of the other. Provisional trust was not really an option in the early days of the space program, but applications are arising today that make provisional trust possible.

Uber, for instance, is a ridesharing platform that pairs drivers of private cars with passengers looking for a ride. Uber can locate a passenger and find the nearest available car. The app also offers information about the driver and details about the car. It certifies that others have ridden in a particular car, that they were safe and comfortable, that they gave the driver a good rating, and that the company has done some level of filtering. This technology represents a new model of community-based provisional trust, created only for the period during which a user needs the service. It is crowdsourced certification that requires very little overhead and infrastructure, and that is highly adaptive to change.

By comparison, a traditional taxi system is much more like the ISS program. It operates in a hierarchical command-and-control structure. There is strict licensing, regulation, and tests. Trust is established not by other users but by the presence of a taxi permit. This is the old tried-and-true model of creating trust, institutional trust, which requires a great deal of infrastructure and overhead, takes a long time to develop, is very inflexible, and is not very open to innovation.

Airbnb is an application similar to Uber, used to locate a room or home to rent. Today, random people are going into random people's houses, spending the night, and comfortably trusting each other. Technology provides built in checks and balances that allow this trust to form. And as with accepting a ride through Uber, the trust between those two parties isn't long-term; it has to last only for as long as one is staying in the other's home.

Airbnb and Uber have become successful because both have

mechanisms to form trust without elaborate processes or contracts. Although bugs are still being worked out, these approaches are showing great promise, with very powerful potential applications to problem solving. Both platforms use a combination of platform-initiated checks and an online community to provide peer-to-peer ratings. These platforms and many more like them rely on virtual communities to make trust building work. When many people review and comment on a particular room for rent or an Uber driver, those evaluations start to become statistically accurate. The driver or homeowner has demonstrated a track record of living up to agreements, and the collective wisdom of the crowd can point to a high level of dependability. This is similar to Duolingo's use of beginning language students to provide translations or ReCAPTCHA's ability to crowdsource the accuracy of book scans.

COMMUNITY-BASED TRUST

These examples relate to personal trust, but there are countless similar examples of communities that form online for a specific purpose and operate in a coordinated way for the greater good. Wikipedia, for instance, was built on the premise that people enjoy interacting within a community, which in the case of Wikipedia, is a global village documenting human knowledge. As is the case with hackathons, very few people work to add knowledge and keep Wikipedia updated out of a sense of charity. Instead, they do it because it's interesting, fun, and community oriented. And like Uber and Airbnb, Wikipedia employs a model that engages the trust of the community to help keep postings accurate and up to date.

There also is a community-based trust model that deals with geographic data, to help keep platforms like Google Maps up to date in the face of changing street names or geography. Many parts of the developing world lack accurately recorded information on which to base maps, because survey data is incomplete. In such cases, only the people who live in an area know the area to a fine level of detail.

In response to this constantly changing environment, Google developed a product called Map Maker. Now, in many cities around the world, information is captured, corrected, and encoded by the citizens who live there. The resulting map is always current and reliable. Once the changes are made and are verified by other users, Google Maps is updated, and the rest of the world can use that updated information. It's sort of a Wikipedia for maps. The wisdom of the crowd keeps the map accurate and updated.

Many of the actions taken in online communities such as Wikipedia or Google Map Maker build trust through demonstrated action. The community tracks not just how many people submit a given change but also who submits the change and how much the system "trusts" that person. In this case, trust is built by demonstrated reliability, and it is built without having a contract in place. It is earned simply through the data a user provides and the value that user adds to the system. When someone has built a reputation for providing good, valuable, accurate information, people begin to trust that new information from that user is accurate and sincere, and this trust can ripple out and extend to other people. This kind of trust happens organically; it is not negotiated or contracted. This is very different from the kind of trust that was required to build the ISS or that would be required for our imaginary ISW, but it is enabled by today's technology.

GIFT ECONOMY

Online communities can build the level of trust required for true collaboration, and they have tremendous potential to bring large numbers of people together. Social media enables people to build trust with their followers, to feel a real connection, and to build a level of trust even though they never meet face-to-face.

Lily Cole is a successful model, actress, and social entrepreneur—though her real claim to fame, in my book, is that as an undergrad at Cambridge she interviewed Stephen Hawking for her thesis paper. Cole founded impossible.com in response to a question that perplexed her: Why do societies collapse in the wake of an economic crisis? "We still have the same amount of resources and skills as we had before the crisis," she said, "but somehow we lost the means by which to get things done. Isn't it problematic that we are so dependent on the one economic structure we've created now?" That question turned into the idea to use technology to enable a gift economy, in which the currency is not money but thank yous.

Impossible is eloquently simple: people post wishes, and others who can, fulfill them. Maybe someone wants someone to practice Spanish with, or has a broken shelf that needs fixing. Other people who have the time and necessary skill fulfill those wishes. No money is exchanged and there's no expectation that the recipient owes the doer anything. It's a simple, one-directional act of goodwill.

Cole named her endeavor Impossible because people told her the idea wouldn't work. But her philosophy and premise in creating Impossible is that competitive structures create antagonisms between individuals, which prevents them from building relationships. "I think that's a very lonely, alienated way for human beings to coexist with one another." Professor Yunus has also equated alienation with lack of relation: "We tend to live in isolation in our own little island, and that creates a problem because I become smaller than I am. If we can connect with each other, we can become bigger than we are."

If Wikipedia is a gift economy on an intellectual level, Impossible is a gift economy on a socioeconomic level. Like the Space Apps Challenge or hackathons in general, these systems are based on the premise that if you give people the tools to help in their communities and in society, they will. This is an important piece of the puzzle in tackling the challenges facing our world. There are vast untapped resources of skills, talent, and intellect that are just now being connected and integrated into our global society.

Cole's goal with Impossible is to use worm's eye details to affect the orbital perspective, or as she put it, "All change begins with yourself." The Impossible platform enables the person-to-person, peer-to-peer exchange that creates a small microsystem of goodness. And if you look at that from the bigger picture, it can change society. When people shift their attitudes toward each other, it will eventually lead to a shift in how countries relate to each other. "If you look at some of the most successful initiatives in the last twenty-five to thirty years," Cole said, "historically they are open systems." She pointed out that the Internet began as a free, open network that people could build on, and that applications such as Wikipedia probably owe their success to their open source nature. "I think the success of cooperative models has been shown to be right."

Wikipedia founder Jimmy Wales believes that the gift economy is something that we're just starting to understand. He pointed to the numerous websites and businesses that are built upon open source software and how this is transforming the world. "And we've just begun to scratch the surface, because of the tools we have today to bring people together. In the past, you really couldn't find one hundred people who wanted to do something. It would be too expensive. What would you do, send letters to everyone?"

BUILDING RELATIONSHIPS FROM A DISTANCE

Our imaginary Project X is shaping up to be an informal, open source, organic, global, community-led initiative. Nonetheless, there still will probably be times where good, old-fashioned, relationship-built trust will be needed to ensure that things run smoothly. During the early days of the U.S.–Russian partnership in space, this was facilitated through face-to-face meetings and even social gatherings among crews and their families. This face-to-face approach was necessary, given the technology of the time. Today, however, information technology allows for actions and agreements to take place between people at a distance. It also is significantly faster and much less formal. Virtual meetings can save a lot of energy, money, and time for all participants and can overcome time zone or geographic distances.

Google's Chief Evangelist for Brand Marketing Gopi Kallayil firmly believes that online communities can build the level of trust required for true collaboration, and although this cannot entirely replace face-to-face meeting and interaction, he believes that's not really the intention. "The intention is to allow people to come together and establish at least a certain level of trust very quickly, across borders and geography, and without the limitations forced by current geopolitics . . . and start operating on some productivity that would not have been possible otherwise." Platforms such as Twitter, Facebook, and Google Plus allow people who know each other to stay in touch, but they also can connect people who have never physically met.

I have experienced this firsthand, having made online connections that eventually became real relationships. For instance, in March 2012, I met Daria Musk and her music partner Ram Rich for the first time. However, when I finally met them in person we were already close friends, because the three of us had become acquainted on Google Plus. As Daria put it, "It turns that first handshake into a hug."

For Daria, this was not a new experience. She has been connecting with strangers online since 2011, when her brother told her about a new website called Google Plus. At the time, twenty-two-year-old Daria was beginning to feel frustrated with the way her career was shaping up. On a whim, she decided to use the video hangout feature on Google Plus and open it up to the public. If anyone showed up, she'd play for them. The hangout could accommodate up to ten people, she said, "and because I'd been playing empty clubs, ten people sounded pretty good to me. Ten people sounded like a show. I didn't expect it to be anything. I just expected it to be another empty room. But in that moment when it happened, I inadvertently found myself standing on a global stage. I was singing to people on different continents, and after the first song I had this global applause." Suddenly she was singing for a man from Buenos Aires, a woman from Texas, a man and his two daughters from Australia. "It was already Sunday morning [in Australia] and it was Saturday night where I was." A man from Norway stayed for the entire six-hour concert, and Daria watched the sun rise in Norway through his window. At that moment, Daria's career made an abrupt course correction, and she took a global following in the millions along with her.

The experience has led Daria to some interesting observations about the possibility of building real relationships online. In some ways, she actually connected better with her globally distributed online audience than with New York City audiences in dark bars and nightclubs. She was able to see people's faces as they reacted to her songs, and she was able to connect personally with her listeners. In the process, Daria formed many deep friendships within her online community.

Since that first hangout concert, Daria has gone on to play many more interactive concerts on her global stage. She even helped change the way one of the biggest companies in the world does business. Seeing Daria's online success exploding, Google flew her out to the Googleplex in Mountain View, California, and she helped Google design Hangouts on Air, which allows hangouts to be live streamed to the public while being recorded and automatically uploaded to YouTube.

Daria's experience is no longer unique, and she believes this trend has great possibility for addressing the world's problems. "If you don't have millions of dollars at your disposal to bring people together to solve global problems, you can at least start with making sure that everybody's invited to the conversation, that everybody can be a part of it from wherever they are, in the comfort of their own world. If we want to survive, we all have to work together, and we have to know each other to do that." Today, anyone with a computer or a phone and an Internet connection can be part of the conversation, and in Daria's experience, this is no less real or useful than traditional face-to-face forums. "If even [a musical performance] can trickle through and translate through the Internet, through the algorithms, and get through to someone on the other side of the world and have them feel something, then I think a simple conversation is going to work just fine."

Whether it's provisional trust, community-based trust, or trust formed through online, virtual interaction, trust can form a basis for global collaboration. There also are efforts like ReCAPTCHA, in which millions of people are helping to digitize books without even necessarily knowing they're doing it. If you don't know you're collaborating, no trust is required. Trust-based communities may be best suited for some challenges, whereas passive collaborative platforms may appropriately address others.

Whether or not you believe that it is possible to develop real trust-filled relationships online, I think it's reasonable to consider Willow Brugh's view: "I think we are able to open the door to build trust more quickly and to be more aware and accepting through online interaction. I think we're also able—and this is the key point—to expand the *web of trust,* where I might not trust you directly but I have met someone that you know and trust them explicitly, and therefore I'll trust you as well." I am really fascinated by this possibility, and we have only begun to scratch the surface of what's possible when we connect seven billion problem solvers—and then connect those problem solvers with needed data and tools.

UNITY NODE

The long voyage to the creation of the International Space Station that began with the U.S.–Russian collaboration was finally realized on December 6, 1998, when NASA astronaut Nancy Currie maneuvered Space Shuttle *Endeavor*'s robotic arm toward the Zarya module that had been launched three weeks earlier from the Baikonur Cosmodrome in Kazakhstan. Zarya (заря) is the Russian word for "sunrise," and it was given to the module to signify the dawn of a new era of international cooperation in space. Nancy slowly maneuvered the end of the robotic arm over the top of Zarya's grapple target, squeezed the trigger on the controller, causing snare wires to close, and captured the 42,000-pound module. Following this orbital ballet, Zarya was mated with the 26,500-pound Unity module from *Endeavor*'s payload bay, and the first step of a twelve-year space construction project began in earnest.

Unity, also known as Node 1, was designed, constructed, launched, and assembled to be the central connector of the entire ISS. Through Unity pass all the lines of communication, data, electricity, life support, and everything else needed to convert a complicated series of modules, solar arrays, radiators, structural elements, and other equipment into a single functional facility that can support life.

Overcoming the challenges facing our global society on Spaceship Earth requires a similar Unity Node–like mechanism to connect organizations and efforts. This mechanism should include a continuum of resource and communications lifelines, in a platform that is open source and accessible. An effective collaborative mechanism would pair challenges with solutions and bring together an international team of people and organizations who will work collaboratively, unifying efforts across sectors to address and respond to humanitarian needs, achieving impact with efficiency and speed.

Because there are many efforts to build universal open source platforms, it is critical to align efforts. We need to make sure there's a path to bring users closest to the problem into the system, lowering barriers to entry to ensure that everyone has a voice in the discussion. This Unity Node would be a collaborative community, a community of trust, working together with a shared purpose. As with a hackathon, there would be a philosophy of contribution in which people are focused not solely on doing their assigned job well but also on making a good contribution to furthering the shared, overarching goal. Great ideas, innovations, and breakthroughs can come from anywhere, but there has to a structure to connect the dots and align people and organizations with shared purpose. Technological advancements have made it extremely economical to access those ideas and incorporate them into solutions. There is no reason a citizens' group, nongovernment organization, or social enterprise should go it alone. The establishment of this Unity Node would allow us to create "dashboards of Earth," or a control room for Spaceship Earth, where critical data affecting our global society could be presented in an actionable way.

The idea of creating this type of information system is not new. In the 1960s, renowned inventor and visionary Buckminster Fuller popularized the term "Spaceship Earth," and he devoted his life to identifying and communicating the interconnections between economic, social, and environmental systems to improve life on our planet. Long before environmental stewardship was prevalent in society, Fuller addressed the reality of our interconnectedness and the synergy that could be produced by working together to find ways to stop wasting the planet's natural resources. He also strove to catalyze the collective intelligence required to fully tackle the global challenges facing all of us.

One of Fuller's efforts to communicate and visualize our interconnectedness was his design for the U.S. pavilion for the 1967 World's Fair in Montreal (Expo '67). The pavilion, a twenty-story geodesic dome, was intended to serve as the control room for Spaceship Earth. Fuller envisioned this control room as a comprehensive data visualization projected onto the spherical inner shell of the geodesic dome. Before the Internet and the open data movement, Fuller envisioned technology that would permit anyone in the world to access the most accurate and up-to-date information about the planet. His envisioned control room would reveal the entire history of resource allocation, transportation, climate, and many other variables in a comprehensible form. Policy makers could then use this user-friendly data to forecast and plan.

Fuller envisioned the modeling of huge data sets and computational models that would be applied to forecasting and could vastly increase humanity's ability to plan for and potentially avoid human-made and natural disasters. These data visualizations would be fed by satellite-derived data and would display the dynamic flows of Earth's interconnected systems. The idea behind these data visualizations was to enable humanity to recognize formerly invisible patterns and to allow for vastly improved forecasting and decision making.

Fuller's plans were well in advance of the technology of the time, and therefore his original vision for the Expo '67 pavilion was not fully realized. However, advancements in computing and visualization technologies over the past five decades have made it possible to realize the original purpose of the Expo '67 pavilion.

The nonprofit Buckminster Fuller Institute, which is dedicated to finding innovative solutions to some of humanity's most pressing problems, wants to use the fiftieth anniversary of Expo '67 as an impetus to finally fully realize Fuller's dream of creating a control room for Spaceship Earth. By renovating and incorporating modern technology into the fifty-year-old geodesic dome, which is currently the Biosphère Environmental Museum, it hopes to create a global hub for collaborative international problem solving. This tool would be made completely open to all, so that it can be of value to society as a whole. As a result, according to Elizabeth Thompson, executive director of the Buckminster Fuller Institute, "planetary planning could be truly democratized, and changing the course of Spaceship Earth can become a cooperative human effort, rather than a task relegated to policy makers."

Fuller envisioned this control room as a global optimization tool. Today, we have the technology to fulfill Fuller's vision. But technology is not enough. Willow Brugh, who helped build the Space Apps Challenge into what is today, said, "A pitfall to me is thinking that technology is a silver bullet, thinking that technology is going to solve this. We often think technology is a Band-Aid we're going to slap on something and everything else will be fixed. It can help us augment the things we're already working on. It can't surpass or supplant what's already there. It can reopen existing conversations; it can help us bring more equality in the world. We know how to talk to each other—we just need to learn how to listen."

Once we are listening to each other, once we are communicating, we need commitment to the collaboration required to fulfill the vision of global optimization of resources and efforts. We must create and support the expectation that humanitarian and development projects (whether run by governments, nongovernment organizations, multinationals, or anyone else) will share their data in meaningful, user-friendly ways to enable, when appropriate, working together seamlessly toward common goals that benefit the global social body. Funders could even encourage and reward this noncompetitive behavior by tying funding to data sharing.

In large part, this effort will also require a commitment to creating a coordinating body that unites the various platforms that are already trying to share data for social and environmental good. Once such a commitment is made, we can use present technology to connect existing stores of data and provide seamless access. Just as the Unity Node on the ISS serves as a conduit of information rather than a central data storage repository, and just as SERVIR provides a conduit to act on big picture data here on the ground, data providers could link the data sets they feel comfortable sharing with the global community. As the value of data sharing becomes better understood and more accepted, providers likely would be willing to share increasing amounts of data. Once this central data sharing tool is created, data visualization hubs such as the control room of Earth envisioned by Fuller could be created, both physically and virtually, all around the world.

Establishing this level of collaboration is of paramount importance because, in the big picture and over the long term, we are working toward our own long-term survival. Perhaps nothing makes this clearer than the 1990 photograph of Earth taken by Voyager 1 from a distance of about 3.7 billion miles. The photograph shows the Earth as a blue-white speck, contained in a fraction of one pixel. Carl Sagan, who had requested that NASA take the photo, said, "For the moment, the Earth is where we make our stand. . . . There is perhaps no better demonstration of the folly of human conceits than this distant image of our tiny world. To me, it underscores our responsibility to deal more kindly with one another, and to preserve and cherish the pale blue dot, the only home we've ever known."[1]

CALL TO ACTION

This book has highlighted just a few of the amazing things that are happening in our global community and a few of the amazing people who are adopting the orbital perspective, stepping outside of their comfort zones, and making a difference—people who, while working with others, not only are sharing resources and information but also are sharing or foregoing credit for their work. These success stories and successful initiatives are reason for hope and optimism.

But this optimism requires continued collaboration and action. Dialogue is great—even, as we've seen, central—but what we need at this moment in history is positive, disruptive action that leads to exponential good, to exponential progress toward solutions to the challenges we face. We need to stop thinking linearly and embrace exponential change—change that comes from being open to innovative partnerships and solutions. This is true innovation.

Our ability to solve our world's greatest challenges is directly proportional to the extent that each of us, on a person-to-person basis, commits to contributing to progress and follows through on that commitment, no matter how large or how small. The contribution a person makes, and his or her response to this call to action, is a personal decision. For me, it involves continuing to work in the development field and continuing to advocate for the pieces of the puzzle that I feel are critical, such as social business, social enterprise, and improved corporate social responsibility practices. It also involves looking for ways to apply the orbital perspective to existing and developing conflict resolution efforts. I am convinced that, rather than exclusively putting time and energy into projects I am personally involved with, I can make a much bigger impact by helping to propel the good that others are doing.

To that end, I, along with an extremely talented group, have launched an effort called Impact CoLab.[2] The goal of Impact CoLab is to use social media and digital marketing to create global grassroots movements to help propel and scale the good created by purpose-driven efforts around the world. The objective is to amplify the impact, reach, and longevity of the efforts of those companies, organizations, and individuals striving to be at the forefront of social and environmental change.

I also am committed to helping as many people as possible have their own orbital perspective experience. To accomplish this, I am working with a powerful team of visionaries devoted to the democratization of space. Our first bold initiative is the creation of Star Harbor Space Training Academy, which will provide all the training necessary for upcoming private spaceflight participants and the crews that will fly them.[3] Our hope is also to provide everyone with a powerful opportunity to experience the orbital perspective, even if they never leave the ground. Human operations in low Earth orbit are no longer the sole domain of large government agencies, and with a new and exciting commercial human spaceflight industry blossoming, it's a very exciting time in history.

My greatest passion and prime focus will be to continue bringing people together to work collaboratively toward solving the problems facing us all. Specifically, I want to continue those efforts that are following the Unity Node model and are working toward creating universal, open source, collaborative platforms. I am passionate about this because I see no larger contribution than to help multiply the good that everyone else is contributing, by helping them to be more effective and to make best use of available resources.

As I said in the beginning of this book, I don't claim to have all the answers. My purpose here is simply to illustrate the need for global collaboration, to highlight some of the things keeping us from working together more effectively, and to start a global conversation around this idea of deeper collaboration. Realizing that the barriers to collaboration are no longer technical but cultural, we must all look for ways to tear down the barriers and overturn the perverse incentives that stifle progress. Before you start something new, though, I recommend that you first research whether the good that you hope to create and the approach that you desire to take is already being done by an existing group. If it is, and it makes sense, join that effort instead of starting a new one.

Though we have been working for decades to find solutions to our global challenges, and we have made great strides, we still have a mountain to climb. Unfortunately, during our past decades of research, we did not collaborate and cooperate efficiently and effectively. But our excuse for that failure no longer exists. Until very recently, we did not have the technological tools to enable true global collaboration among governments, universities, industries, and citizen scientists. We now have this, and thus we have no excuse. Let's set aside our differences and work together toward our common goals.

Nothing is impossible. The future is ours to build. It's a place that we build in what we choose to do in every given moment. The orbital perspective is the linking up of the big picture view and the worm's eye view. It's being on the ground and knowing the details affecting the impoverished people that you're trying to help, while understanding from the big picture perspective (both spatially and temporally) the long-term effect of a particular intervention or development project, on a global—not just a local—scale. To be truly effective and to provide lasting progress, we must synthesize both views.

HOME

In 2011, at the end of our five-and-a-half-month mission on board the International Space Station, I climbed into the Soyuz spacecraft with my crewmates Sasha and Andrei. We closed the hatch, wiggled ourselves into our seats, and strapped in. Shoehorned into the spacecraft with our knees in our chests, we undocked and slowly backed away from the space station. Seated in the right seat, I strained to keep the space station in view out the right window for as long as possible. I realized that in all likelihood I would never see this magnificent sight again. After undocking, we did a couple of laps around the planet, and as we passed the southern tip of South America, we fishtailed the spacecraft around to point our engines backward. As we did this maneuver, a crescent moon passed gently across my window. On command, at precisely the right moment, we fired the engines of the spacecraft just long enough to slow us enough to enter the upper levels of the atmosphere.

The large sledgehammer bang of explosive bolts separated our spacecraft into three distinct pieces—the bulbous orbital module on the front of the spacecraft, the engine compartment on the aft, and the center descent module, which contained the three of us and the only heat shield. I watched as debris from the separation tumbled out from us in all directions. If all went as planned, the part we were in would be the only part to survive the fiery reentry.

As we entered the atmosphere, the black sky started to turn pink as the temperature increased around our spacecraft. I watched sporadic sparks shoot past the window and rapidly increase in intensity until it resembled the fountain of sparks seen in some fireworks displays. As our altitude decreased, the sense of speed increased rapidly, in spite of what appeared to be a very steep descent. I thought, "There goes Africa" as we rapidly passed across the continent. Soon the sparks became flames and my window turned black with soot. What followed was a very violent ride through the atmosphere. At one point, Andrei, who was in the left seat, yelled out in Russian, "It's just like an American amusement park ride!" But this was far more violent than any ride I had been on.

When we were finally stabilized under the parachute, a layer of film was jettisoned from our windows and we could see out again. I watched through the mirror on my wrist as the ground rose up to meet us. We slammed into the ground, considerably harder than I had anticipated, and I thought, "I'm glad that's over," not knowing that we had bounced and still had a few more "landings" to go. People had told me that, because my vestibular system was all out of whack, I would feel like I was tumbling when I landed, and I noticed this tumbling feeling right away. But when I started seeing debris flying all around the cockpit, reminiscent of those slow-motion rollover crash test videos, I realized that we actually *were* tumbling.

When we came to rest, the spacecraft was lying on its side with me on the bottom, Sasha wedged into my left shoulder, and Andrei wedged into his, nice and cozy with all the other debris that had been dislodged during the impact. As we hung there in our seats, waiting for the rescue forces to arrive, I looked out the window that was now pointed at the ground. I saw a rock, a flower, and a blade of grass, and I thought, "I'm home." Then the significance of that thought dawned on me. I was home, but I was in Kazakhstan. In that moment, I realized in a very profound way that my home is no longer only Houston, Texas, where I currently live; it is Earth.

I left the ISS in 2011 with a call to action, a call to spread this orbital perspective. I want to continue to spread this message of cooperation around the world and to continue working to improve life on Earth, to continue to look for real, sustainable solutions that have the greatest long-term effect. Like an asteroid deflection mission, in which a slight force early enough and far enough out can prevent an extinction-causing asteroid from hitting Earth, I want, through my small actions, to nudge the trajectory of our global society. I want to change perceptions, overturn stereotypes, and counter the demonization of whole populations of people. I want to travel to countries around the world (any country, regardless of ideology or political alliance) and share this message of hope.

If there's one thing I've learned in my travels around the world, it's that people are people. There are more things that we share in common than things that separate us. We may be born in different countries, belong to different generations, speak different languages, but we are one species, with the same hopes and dreams for our children, for our communities, and for our planet.

NOTES

Introduction: A Shift in Perspective

1. Robert F. Kennedy, "Day of Affirmation Address" (speech delivered at the University of Capetown, Capetown, South Africa, June 6, 1966).

Chapter 1: Humanity's Home in the Heavens

1. Committee on Aeronautical and Space Sciences, *Soviet Space Programs, 1971–75* (Washington, DC: U.S. Government Printing Office, 1976).

2. Susan Eisenhower, *Partners in Space: U.S.–Russian Cooperation after the Cold War* (Gettysburg, PA: Eisenhower Institute, 2004).

3. All Daniel Goldin recollections here are from Eisenhower, *Partners in Space,* 152.

4. John F. Kennedy, "Address at Rice University on the Nation's Space Effort" (speech delivered at Rice University, Houston, TX, September 12, 1962).

Chapter 3: Lessons in Collaboration from the ISS Program

1. In a garden near the Baikonur Cosmodrome, a tree is planted for every person since Yuri Gagarin who has launched to space from the facility. It is a quiet, peaceful, and symbolic place, steeped in history, and it was a great honor to join my Russian crewmates in planting our trees a few days before our 2011 launch.

Chapter 4: One Moment in Space

1. Wasfia Nazreen, "Bangladesh on Seven Summits," Fragile Oasis blog post, February 12, 2014, http://www.fragileoasis.org/blog/2014/2/bangladesh-on-seven-summits.

Chapter 5: The Orbital Perspective

1. Frank White, *The Overview Effect: Space Exploration and Human Evolution*, 2nd ed. (Reston, VA: American Institute of Aeronautics and Astronautics, 1998).

2. A number of other human-made borders can be seen clearly from space, including the border between North and South Korea, the United States–Mexico border, the border around Israel, and others.

3. The Eucharist is the consecrated, unleavened wafer used in Holy Communion. A monstrance is a vessel used to display the Eucharist. It contains a circular center to store the host and usually has rays emanating from the center, resembling a depiction of the sun.

4. Microfinance involves providing small loans to impoverished borrowers, which they can use to start businesses.

5. Muhammad Yunus, interview by Wayne Visser, Cambridge Programme for Sustainability Leadership, 2008, http://www.youtube .com/watch?v=f6eOKVuAw7g.

6. The B612 Foundation, founded by retired astronauts Ed Lu and Rusty Schweickart and Dutch astrophysicist Piet Hut, is a great organization that is attempting to launch the Sentinel asteroid early warning system. http://sentinelmission.org.

7. I am working on a campaign called Vision Zero to reduce the number of worldwide traffic fatalities to zero by 2030.

8. Robert F. Kennedy, "Day of Affirmation Address."

Chapter 6: The Key Is "We"

1. Emmanuel Jal, *War Child: A Child Soldier's Story* (New York: St. Martin's Press, 2009), 256.

2. Amanda Lindhout and Sara Corbett, *A House in the Sky* (New York: Scribner, 2013), 293.

3. Ibid., 366.

4. Ibid., 364.

Chapter 7: Camp Hope

1. Faaiza Rashid, Amy C. Edmondson, and Herman B. Leonard, "Leadership Lessons from the Chilean Mine Rescue," *Harvard Business Review*, July–August 2013.

2. Albert W. Holland, interview by Rebecca Wright, Houston, TX, April 25, 2011. NASA Chilean Miners Rescue Oral History Project,

http://www.jsc.nasa.gov/history/oral_histories/CMR/HollandAW/
HollandAW_4-25-11.htm.

3. Ibid.

4. Ibid.

5. Ibid.

6. Ibid.

7. Ibid.

8. Ibid.

9. Ibid.

10. James D. Polk, interview by Rebecca Wright, Houston, TX, May 27, 2011. NASA Chilean Miners Rescue Oral History Project, http://www .jsc.nasa.gov/history/oral_histories/CMR/PolkJD/PolkJD_5-27-11.htm.

11. Ibid.

12. Ibid.

13. BBC News, "Chile Mine Rescue: Last Miner Luis Urzua Reaches Surface." YouTube video, 7:40. Posted October 13, 2010. https://www .youtube.com/watch?v=hRdy2gx8_Kc.

14. Rory Carroll and Jonathan Franklin, "Chile Miners: Rescued Foreman Luis Urzúa's First Interview," *The Guardian,* October 14, 2010.

15. "NASA Astronauts Speak to Chilean Miners," The History Channel website, http://www.history.com/speeches/nasa-astronauts-speak-to-chilean-miners#nasa-astronauts-speak-to-chilean-miners.

16. Polk, interview by Rebecca Wright.

17. Michael Useem, Rodrigo Jordán, and Matko Koljatic, "How to Lead during a Crisis: Lessons from the Rescue of the Chilean Miners," *MIT Sloan Management Review,* Fall 2011.

Chapter 8: Arrested Development

1. Founding members of the Manna Energy Foundation were Ron and Carmel Garan, Chris and Nicole Stott, Michael Potter, John Garan, and Victor Bernstein.

2. Evan Alexander Thomas, "Proving Global Impact: How the Integration of Remotely Reporting Sensors in Water Projects May Demonstrate and Enhance Positive Change," *Global Water Forum,* December 1, 2013, http://www.globalwaterforum.org/2013/12/01/ improving-global-impact-how-the-integration-of-remotely-reporting-sensors-in-water-projects-may-demonstrate-and-enhance-positive-change.

3. Tina Rosenberg, "To Maintain Water Pumps, It Takes More Than a Village," *New York Times,* December 13, 2011, http://opinionator.blogs

.nytimes.com/2011/12/13/to-maintain-water-pumps-it-takes-more-than
-a-village.

4. Improve International, *Statistics on Water Point Failures,* http://
improveinternational.wordpress.com/handy-resources/sad-stats.

Chapter 9: Mass Collaboration

1. Discussions took place at an "unconference" to help coordinate
disaster and crisis response, called Crisis Camp. Panelists included Phil
Dixon and Jeff Martin from Google, Jeremy Johnstone from Yahoo, and
Patrick Svenburg from Microsoft, with Greg Elin from the Sunlight
Foundation moderating.

2. Luis von Ahn, "Massive-Scale Online Collaboration." Filmed April
2011. TED video, 16:39. http://www.ted.com/talks/luis_von_ahn_massive
_scale_online_collaboration?language=en.

Conclusion: A Web of Trust

1. Carl Sagan, *Pale Blue Dot: A Vision of the Human Future in Space*
(New York: Random House, 1994), 7.

2. Founding members of Impact CoLab are Ron Garan, Elyse David,
Krishan Arora, Ness Knight, Daria Musk, Dan Cook, and Ali Llewellyn.

3. Star Harbor Space Training Academy is a project conducted
through Space Development Ventures. The founding CEO is Maraia
Hoffman and founding members include Shubham Garg, Tim
Bailey, Alan Ladwig, Robert Ward, Jacob Hockett, Luis Marquez, and
retired astronauts Leland Melvin and Ron Garan. The effort is also in
partnership with Crux Partners, Impact CoLab, and Fragile Oasis.

ACKNOWLEDGMENTS

In doing the research for this book I followed the same philosophy I use in photography: if you take enough pictures, eventually you'll get a few that are great. In pulling together the concepts described in this book, I asked countless questions of some really smart people and eventually some of these questions turned out to be the correct questions. Most of the people I queried are friends of mine, and many of those friends became friends before we ever met in person. Inclusion of most, if not all, of the voices in this book was made possible because of the exponential increases in our global interconnectivity. My task in writing this book was to connect the dots, to link answers to the questions I asked, and to combine that with my own thirty-plus-year experience working in collaborative systems.

Special thanks to Neal Maillet and all the wonderful and talented people at Berrett-Koehler Publishers for taking a chance on an unknown writer. I am very thankful that my first book is being produced by a publisher that exudes the orbital perspective through its socially and environmentally conscious business practices and its goal of "Creating a world that works for all." Thanks also to Jeevan Sivasubramaniam for his guidance, and to the reviewers of my first draft: Nic Albert, Onnesha Roychoudhuri, Rich Landry, and Ellyn Kerr. Special thanks to the amazing Todd Manza for his expert editing, and for using a scalpel and not a hatchet to trim the book down to its current form.

I want to personally thank those friends, colleagues, and my personal heroes who selflessly provided their time and patience and shared their unique and powerful perspectives. Special thanks to my partner in Fragile Oasis, Executive Producer Elyse David, and my colleagues at Impact CoLab, Manna Energy Foundation, Manna Energy Ltd., and Star Harbor Space Training Academy. Thanks to NASA for giving me the opportunity to experience the orbital perspective, and to Professor Muhammad Yunus for teaching me the importance of synthesizing a worm's eye view. Thanks to astronauts and cosmonauts Mike Foale, Mike Barratt, Tom Marshburn, Leroy Chiao, Koichi Wakata, and Ellen Baker, Sergei Krikolev, Vladimir Titov, and NASA Administrator Charlie Bolden for your contributions to the book. Also, a big thanks to retired astronaut General Kevin Chilton for the expert lesson in geopolitics.

Thanks to everyone who has added their voice to this book, including Al Holland, John McBrine, Bill Gerstenmaier, George Abbey, Mike O'Brien, Jeff Manber, Joan Johnson-Freese, Samantha Snabes, former NASA Administrator Mike Griffin, Marcelo Aliaga, Gopi Kallayil, Vic Gundotra, Jimmy Wales, Luis von Ahn, Willow Brugh, Wasfia Nazreen, Elena Maroko, Lily Cole, Evan Thomas, Daria Musk, Nick Skytland, Chris Gerty, Ali Llewellyn, Patrick Svenburg, Stu Gill, Rebecca Wright, Emmanuel Jal, Elizabeth Thompson, Amanda Lindhout, and Dan Irwin.

Thanks to everyone who has provided moral support through my writing journey, including Chris and Nicole Stott, Ness Knight, Dom Dauster, Hans Reitz, Leland Melvin, Amy Moore-Benson, Maraia Hoffman, and Clay Morgan.

Thanks to all those who are helping to spread the message of the orbital perspective. Thanks also to my mother and father, who started and supported me on this journey on and off Spaceship Earth. Last but not least, I want to thank Carmel, Ronnie, Joseph, and Jake.

INDEX

ABOUT THE AUTHOR

Col. Ron Garan (USAF ret.) is a decorated fighter pilot, astronaut, aquanaut, and social entrepreneur. He has logged 178 days in space and 71 million miles in orbit. He is the founder of the non-profit social enterprise incubator Manna Energy Foundation and has worked with the U.S. Agency for International Development. Col. Garan is also the founder of Fragile Oasis (FragileOasis.org), an effort to use the orbital perspective to inspire positive social and environmental action.

Berrett–Koehler
Publishers

A community dedicated to creating
a world that works for all

Visit Our Website: www.bkconnection.com

Read book excerpts, see author videos and Internet movies, read
our authors' blogs, join discussion groups, download book apps, find
out about the BK Affiliate Network, browse subject-area libraries of
books, get special discounts, and more!

Subscribe to Our Free E-Newsletter, the *BK Communiqué*

Be the first to hear about new publications, special discount offers,
exclusive articles, news about bestsellers, and more! Get on the list
for our free e-newsletter by going to **www.bkconnection.com**.

Get Quantity Discounts

Berrett-Koehler books are available at quantity discounts for orders
of ten or more copies. Please call us toll-free at (800) 929-2929 or
email us at **bkp.orders@aidcvt.com**.

Join the BK Community

BKcommunity.com is a virtual meeting place where people from
around the world can engage with kindred spirits to create a world
that works for all. **BKcommunity.com** members may create their own
profiles, blog, start and participate in forums and discussion groups,
post photos and videos, answer surveys, announce and register for
upcoming events, and chat with others online in real time. Please join
the conversation!